JN297131

# 英語で学ぶ
# 土 質 力 学 (2)
— 力学的性質編 —

Soil Mechanics (2)
— Mechanical Properties —

酒井　俊典
勝山　邦久
**Md. Zakaria Hossain**　共著
**Laura J. Pyrak-Nolte**

コロナ社

# まえがき

　皆さんが土を手に取ってみると，土は，土粒子の粒だけでできているのではなく，土粒子の間に隙間（間隙）があり，その間隙に水や空気を含んだ三つの要素からできていることがわかる。この土は，もともと岩石などが風化してできたものであり，鉄やコンクリートのような人間が作り出した人工的な材料ではなく，その性質は，周りの自然環境や生成の過程によって大きく異なる。このため，不均一で多種多様な特性を示し，複雑でやっかいな材料である。

　現在まで，人間はこの土を利用して水路，道路，トンネルといったさまざまな構造物を造り，生活を便利にしてきた。土を材料として使い，安全な構造物を設計・施工する場合，土粒子の大きさやその分布状態，あるいは間隙中の水分あるいは空気の存在の仕方によって，特性が大きく異なることを理解しておくことが重要である。もし，水路を造ろうとした場合，これら土の特性を十分理解しないで掘削を行うと，掘削中に崩れてきて危険な状況となるかもしれない。また，道路を建設した後，しばらくして道路が崩れることがあるかもしれない。昔の人は，こういった危険な状況をいくつも経験し，これらの経験をたくさん蓄積することで，さらに安全な構造物を造ることにつなげてきた。

　土質力学はこのような複雑な土を扱ううえで，経験だけに頼るのではなく，その特性を物理的，力学的，化学的なさまざまな面から体系化しようとしているものである。しかし，自然材料である複雑でやっかいな土を扱う土質力学の体系は，場合に応じて弾性体，塑性体，粘性体，あるいは粒状体など種々な考え方を取り入れる必要があり，まだまだ完全には体系化されておらず，ある意味発展途上の学問であるとも言える。このため，逆に言えば土質力学は研究する部分が多く残された，魅力ある分野であるとも考えることができる。

　現在インターネットの普及などによって国際化が加速し，各分野で国際基準

が取り入れられている。このように国際化が進み海外からの留学生も増加している現在，英語の必要性はさらに増し，土質力学の分野においても日本国内で英語による講義が行われている例も見られるようである。土質力学において英語を勉強する場合，一般的な辞書には記述されていない専門用語が多くあるため，これら専門的な英語表現を簡単に取り込めることができれば，比較的容易に英語に接することができるのではないかと思う。

本書は，土木構造物の設計・施工において必要な基本的な事項を中心に，日本語と英語を対比させながら記述することで，土質力学の学び初めの人でも容易に英語に取り組めるよう配慮して執筆を行った。執筆にあたり英語のチェックは，地球環境科学を専門とし，世界中を飛び回っているLaura先生にお願いした。また，息抜きに土質力学に関係のある人物の話などもコーヒーブレークのように挿入し，土質力学の歴史的な側面も知ってもらえればと思っている。本書が，複雑でやっかいな土質力学を学習しようとする日本人学生だけではなく，日本に来ている留学生や，海外で働く技術者など，多くの人々に利用していただき，少しでも土質力学に興味を持っていただければ光栄である。

本書は，基本的物理量，土中の水，圧密，地盤内応力といった土の基本的性質に関連する事項を中心にまとめた(1)巻の続巻として，応力，土のせん断，土圧，支持力，斜面安定といった土の力学的性質に関連する事項を中心にまとめた(2)巻として執筆・編集した。この(1)・(2)巻を通して土質力学全般についての学習ができるようになっている。

最後に，本書の出版に際してお世話になった株式会社コロナ社に感謝申し上げます。

2010年7月

酒井 俊典

# CONTENTS

## Chapter 6　Stress and Strain

6.1　Normal Stress ································································2
6.2　Strain ············································································4
6.3　Shear Stress ·································································6
6.4　Shear Strain ··································································6
6.5　Stress-Strain Relationships ·········································8

## Chapter 7　Mohr's Stress Circle

7.1　Mohr's Stress Circle for Principal Stress ················14
7.2　Conjugate Relationships of Shear Stresses ············22
7.3　Mohr's Stress Circle Considering Shear Stresses ···24

## Chapter 8　Shearing of Soil

8.1　Shear Strength of Soil················································34
8.2　Coulomb's Failure Criterion ·····································36
8.3　Mohr-Coulomb's Failure Criterion ·························46
8.4　Shear Properties of Sand and Clay·························54
8.5　Shear Testing Apparatus ·········································58
8.6　Consolidated and Drained Conditions····················72

## Chapter 9　Earth Pressure

9.1　Types of Earth Pressure ············································84
9.2　Rankine's Earth Pressure ··········································90
9.3　Coulomb's Earth Pressure ·········································96

## Chapter 10　Bearing Capacity

10.1　Bearing Capacity Formula ·······································106
10.2　Shallow Foundation ··············································114
10.3　Deep Foundation ·················································120

## Chapter 11　Slope Stability

11.1　Limit Equilibrium Method··········································126
11.2　Stability Analysis Considering Circular Slip Surface ······136

**Appendix**··································································144
**References** ······························································152
**Index** ···································································154

### Soil Mechanics (1)
— Fundamental Properties —

#### Main CONTENTS

1　Definition of Soil　　　　2　Basic Properties of Soil
3　Flow of Water in the Ground　　4　Consolidation
5　Stresses in the Ground

# 目 次

## 6 応力とひずみ

6.1 垂直応力 …………………………………………………… 3
6.2 ひずみ ……………………………………………………… 5
6.3 せん断応力 ………………………………………………… 7
6.4 せん断ひずみ ……………………………………………… 7
6.5 応力-ひずみ関係 …………………………………………… 9

## 7 モールの応力円

7.1 主応力によるモールの応力円 …………………………… 15
7.2 せん断応力の共役関係 …………………………………… 23
7.3 せん断応力を考慮したモールの応力円 ………………… 25

## 8 土のせん断

8.1 土のせん断強度 …………………………………………… 35
8.2 クーロンの破壊基準 ……………………………………… 37
8.3 モール・クーロンの破壊基準 …………………………… 47
8.4 砂と粘土のせん断特性 …………………………………… 55
8.5 せん断試験機 ……………………………………………… 59
8.6 圧密・排水条件 …………………………………………… 73

## 9 土　　圧

9.1 土圧の種類 …………………………………… 85
9.2 ランキン土圧 ………………………………… 91
9.3 クーロン土圧 ………………………………… 97

## 10 支　持　力

10.1 支持力公式 ………………………………… 107
10.2 浅い基礎 …………………………………… 115
10.3 深い基礎 …………………………………… 121

## 11 斜 面 安 定

11.1 極限釣合い法 ……………………………… 127
11.2 円弧すべり面の安定解析 ………………… 137

付　　録 ………………………………………… 144
参 考 文 献 …………………………………… 152
英 和 索 引 …………………………………… 154

### 英語で学ぶ 土質力学（1）
— 基本的性質編 —

#### 主 要 目 次

1　土とは　　　2　土の基本的物理量
3　土中の水　　4　圧　　密
5　地盤内応力

# 本書を学ぶにあたって

　本書は，土質力学の基本的事項を中心にまとめているため，初めて土質力学を学ぶ読者にとっては，始めに日本語の部分を学ぶことで，土質力学に必要な基礎的な内容を理解することができると思う。そのうえで，英語の部分と対比をすることで，土質力学で用いられる専門用語の英語での表現，あるいはその使い方を知ることができる。

　本書は，英語・日本語は直訳にはなっていないが，高校前半の実力で理解することができる平易な英語を使っているので，英語の得意な読者は，最初から英語の部分から読まれてもよいと思う。また，土質力学関係の専門用語を調べる必要があるときの英和・和英辞書として利用することも一案である。

　各節の終わりには内容の理解を深めるため例題を設けており，日本語だけでなく英語の問題として解いていただければと思う。

　付録には三軸圧縮試験の種類，土質試験の種類，ボーリング調査，日本地質概要図，英語の数式の読み方や，土質力学に記号としてよく用いられるギリシャ文字の読み方も添付している。

　国際化が進む中，本書により日本語と英語との対比によって土質力学を学ぶことで，少しでも国際的な意識を持っていただき，外国人とのディスカッション，国際学会やシンポジウムなどでの発表の手助けになれば光栄である。

# Chapter 6  Stress and Strain

## 6.1  Normal Stress

When an external force acts on a body (material), the body deforms producing internal strain and stress (internal force) inside the body. The stress component perpendicular to each plane is called the normal stress. The normal stress can be either tensile or compressive depending on the direction of the stress acting on the body.

When a uniform compressive load $P$ acts on a body of cross-sectional area $A$ as shown in **Figure 6.1**, the normal stress $\sigma$ produced inside the body is compressive and is given by:

$$\sigma = \frac{P}{A} \tag{6.1}$$

Plus and minus signs (+, −) are used to indicate the direction of the normal stress. In structural mechanics, a plus sign (+) is used for tensile stress and a minus sign (−) is used for compressive stress. However, in soil mechanics, a plus sign (+) is used for compressive stress because tensile stress does not developed in soil (i.e., soil cannot support tension). The units of stress are $N/m^2$ or Pa (Pascal).

# 6章 応力とひずみ

## 6.1 垂直応力

物体に外力が作用すると物体は変形し，内部にひずみが生じるとともに応力（内力）が発生する。この応力を垂直応力といい，作用方向によって引張応力と圧縮応力に分けられる。

**図6.1**のような断面積 $A$ の物体に均等に荷重 $P$ を作用させると，物体内部に発生する垂直応力 $\sigma$ は，荷重を断面積で除した

$$\sigma = \frac{P}{A} \tag{6.1}$$

で表され，このときの応力は圧縮応力である。

垂直応力の作用方向による正負（＋，－）は，構造力学では引張応力を正（＋），圧縮応力を負（－）とするのに対し，土質力学では，土は引張りに耐えられないため引張応力が発生しないと考え，圧縮応力を正（＋）として扱う。

垂直応力に用いる単位は，ニュートンを単位面積で除した kN/m² が用いられる。

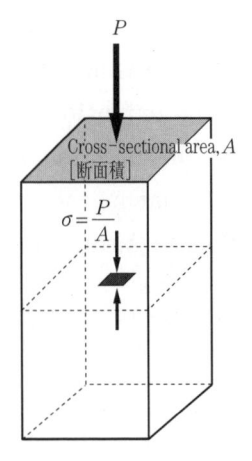

Normal stress is a function of the compressive load $P$ and the cross-sectional area $A$.

**Figure 6.1** Normal stress
**図 6.1** 垂直応力

## 6.2 Strain

A dimensional change in the size and shape of a body occurs when a compressive or tensile stress is applied. For example, as shown in **Figure 6.2**, an applied compressive load, $P$, results in longitudinal shortening and transverse lengthening (or widening). The dimensional change in a sample is quantified by using strain. Strain is the ratio of the altered length, area, or volume relative to its original value. If the original length is $l$ and the altered length is $\Delta l$, then the strain $\varepsilon$ is given by the following equation:

$$\varepsilon = \frac{\Delta l}{l} \qquad (6.2)$$

Strain is non-dimensional and has no units. Sometimes strain is given as a percentage.

If no load is applied in the transverse direction during deformation of a body as shown in Figure 6.2, there is an extension in the transverse and a contraction in the longitudinal direction. The ratio of the extensional or transverse strain ($\varepsilon_h$, that is, the strain normal to the direction of the applied load) to the contraction or axial strain ($\varepsilon_v$, that is, the strain in the direction of the applied load) is called Poisson's ratio $\nu$ and is defined as:

$$\nu = -\frac{\varepsilon_h}{\varepsilon_v} \qquad (6.3)$$

## 6.2 ひ　ず　み

物体に圧縮（引張）力を加えると，物体は縮み（伸び）によって図 6.2 のように変形する。このとき，物体の元の長さ $l$ に対して縮んだ（伸びた）量 $\Delta l$ を，ひずみ $\varepsilon$ といい

$$\varepsilon = \frac{\Delta l}{l} \tag{6.2}$$

で表される。

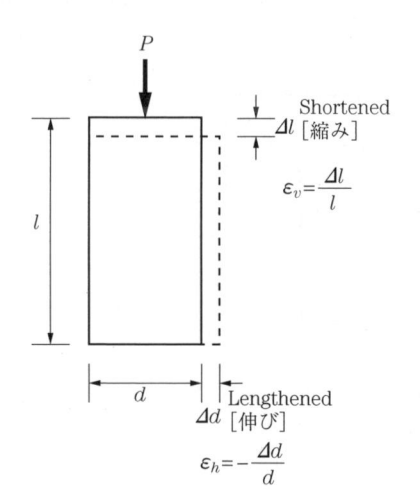

**Figure 6.2** Longitudinal strain $\varepsilon_v$ and transverse strain $\varepsilon_h$
図 6.2　縦ひずみ $\varepsilon_v$ と横ひずみ $\varepsilon_h$

ひずみは単位がなく無次元であり，場合により百分率（％）が用いられる。

2 次元における物体の変形を考えると，図 6.2 に示すように縦方向に物体が縮んだ場合，それと直交する横方向は伸びることとなる。このとき，縦方向のひずみ（縦ひずみ：$\varepsilon_v$）と横方向のひずみ（横ひずみ：$\varepsilon_h$）との関係は，ポアソン比 $\nu$ により

$$\nu = -\frac{\varepsilon_h}{\varepsilon_v} \tag{6.3}$$

で定義される。

## 6.3 Shear Stress

Shear stresses develop in a body when two parallel shear forces $S$ act on the body in opposite directions as shown in **Figure 6.3**. The shear stress $\tau$ is calculated by dividing the shearing force $S$ by the shearing area $A$ as given in Equation (6.4).

$$\tau = \frac{S}{A} \tag{6.4}$$

Like the normal stress, the units for shear stress are N/m² or Pa (Pascal).

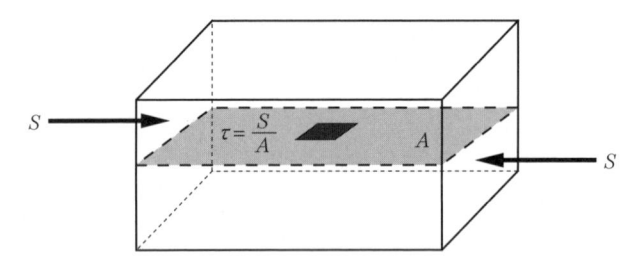

Shear stress $\tau$ from the application of shear forces $S$ over the shearing area $A$

**Figure 6.3**  Shear stress
図 6.3  せん断応力

## 6.4 Shear Strain

For a linear elastic body, when a quadrilateral area ABCD is subjected to a shear force, it deforms to form a parallelepiped A′BCD′ as shown in **Figure 6.4**. The shear strain, $\gamma$, defines the shear deformation for an applied shear force and is given by the angle, $\angle$ABA′(or similarly, $\angle$EFE′. Shear strain is non-dimensional and therefore has no units.

## 6.3 せん断応力

図 6.3 のように，二つの平行で向きが反対のせん断力 $S$ が物体に作用し，物体をずらそうと（せん断）するとき，物体内部にはせん断応力 $\tau$ が発生する。このときせん断応力は，せん断力が作用する面積 $A$ で除した

$$\tau = \frac{S}{A} \tag{6.4}$$

で表される。

せん断応力に用いる単位は，垂直応力と同様，ニュートンを単位面積で除した $kN/m^2$ が用いられる。

## 6.4 せん断ひずみ

弾性体がせん断力を受けると，図 6.4 に示す四角形 ABFE は，平行四辺形 A′BFE′ に変形する。このとき，変形前後の角度 $\angle ABA'$ ($= \angle EFE'$) は，せん断応力の大きさによって変化する量で，これをせん断ひずみ $\gamma$ と定義する。せん断ひずみの単位は，ひずみと同様無次元である。

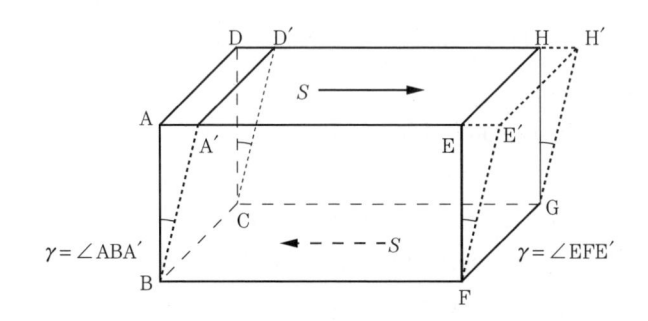

**Figure 6.4** Shear strain
図 6.4　せん断ひずみ

## 6.5 Stress-Strain Relationships

A material is said to be elastic when it returns to its original form upon withdrawal of an applied load. An elastic body follows Hooke's law that gives a linear relationship between stress and strain (**Figure 6.5** on the left). Hooke's laws for normal stress and shear stress are:

$$\sigma = E\varepsilon \tag{6.5}$$
$$\tau = G\gamma \tag{6.6}$$

Here, $E$ is the coefficient of elasticity or Young's modulus (N/m²), $G$ is the coefficient of shear elasticity or shear modulus (N/m²).

(a) Elastic body [弾性体]  (b) Rigid-plastic body [剛塑性体]  (c) Elasto-plastic body [弾塑性体]

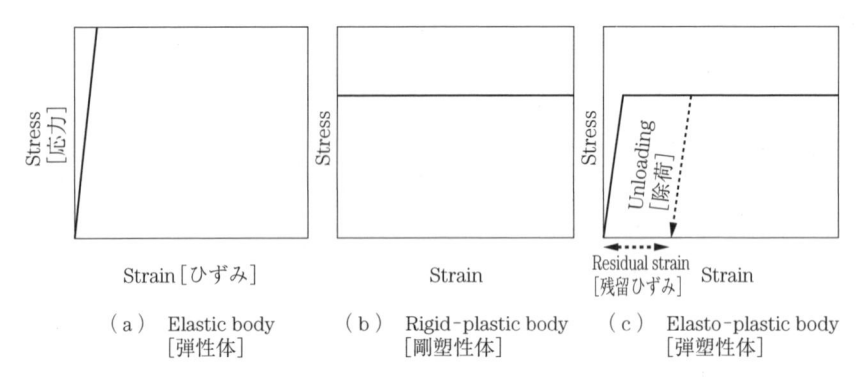

**Figure 6.5** Stress-strain relationships
図 6.5 応力とひずみ関係

For soil, Hooke's Law only applies for very small deformations (Figure 6.5). In general, large deformations occur just after the yield stress when a material begins to deform plastically. Residual strains remain in the soil even when the applied load is removed (Figure 6.5 (a)). Plastic deformation and residual strains need to be considered when working with soil. A soil should be represented as a rigid-plastic body when small elastic deformations are neglected (Figure 6.5 (b)) or as an elasto-plastic body (Figure 6.5 (c)) when

## 6.5 応力-ひずみ関係

物体に作用する荷重を取り去ると元の状態に戻る材料を弾性体という。弾性を示す材料は，応力とひずみが比例関係を示すフックの法則が成り立ち，垂直応力とひずみ，せん断応力とせん断ひずみはそれぞれ

$$\sigma = E\varepsilon \tag{6.5}$$
$$\tau = G\gamma \tag{6.6}$$

で示される。

ここで，$E$ は弾性係数（$kN/m^2$），$G$ はせん断弾性係数（$kN/m^2$）と呼ばれる定数である。

材料が示す応力とひずみとの関係は，図6.5に示す（a）弾性体，（b）剛塑性体，（c）弾塑性体などが考えられる。

ところで，土の弾性変形は変位がごく小さい範囲であり，降伏点を超えると変形が急速に大きくなり，荷重を取り去っても残留ひずみが残る塑性変形が起きる。このため，土を扱う場合，弾性変形を微小として無視する剛塑性体，あるいは弾性・塑性の両方を考慮する弾塑性体の考え方が必要となる。

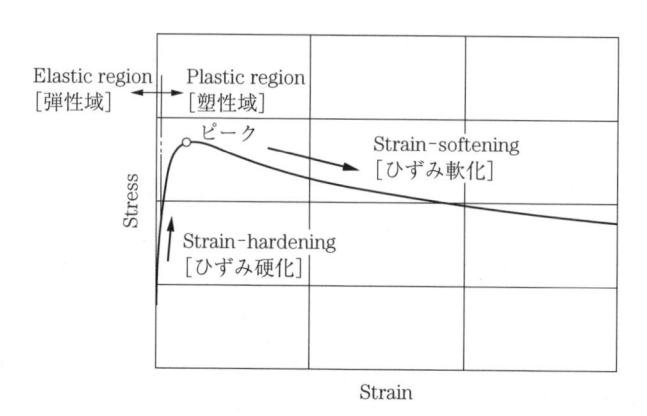

**Figure 6.6** Stress-strain relationship considering strain-softening phenomenon
図6.6 ひずみ軟化を伴う応力とひずみの関係

both the elastic and plastic properties of a soil are considered. In fact, the shear behavior of a soil depends on the compaction and the stress history of the soil. A strain-softening phenomenon can also develop where a decrease in stress results in an increase in strain after a peak stress (**Figure 6.6**). All of these plastic phenomena need to be considered to describe the behavior of soil accurately.

**〖Example 6.1〗** Calculate the stress-strain relationships in a two-dimensional plane-stress condition for an elastic body.

**Solution** The relationships between longitudinal strain $\varepsilon_h$ and transverse strain $\varepsilon_v$ can be written in terms of Poisson's ratio $\nu$ as follows:

$$\varepsilon_h = -\nu\varepsilon_v \tag{6.7}$$

$$\varepsilon_v = -\nu\varepsilon_h \tag{6.8}$$

The transverse strain $\varepsilon_x$ and the longitudinal strain $\varepsilon_y$ in two-dimensions are given by:

$$\varepsilon_x = \varepsilon_h - \nu\varepsilon_v \tag{6.9}$$

$$\varepsilon_y = -\nu\varepsilon_h + \varepsilon_v \tag{6.10}$$

Using Hooke's law, ($\sigma = E\varepsilon$), Equations (6.9) and (6.10) become;

$$\varepsilon_x = \frac{\sigma_x}{E} - \nu\frac{\sigma_y}{E} \tag{6.11}$$

$$\varepsilon_y = -\nu\frac{\sigma_x}{E} + \frac{\sigma_y}{E} \tag{6.12}$$

The shear strain $\gamma_{xy}$ and coefficient of shear elasticity $G$ are given by:

$$\gamma_{xy} = \frac{\tau_{xy}}{G} \tag{6.13}$$

$$G = \frac{E}{2(1+\nu)} \tag{6.14}$$

The following equation is obtained:

$$\gamma_{xy} = \frac{2(1+\nu)}{E}\tau_{xy} \tag{6.15}$$

Using Equations (6.11), (6.12) and (6.15), the full relationship between stress and strain for a two-dimensional system are:

さらに，実際の土のせん断挙動は，地盤の応力履歴や密度などの違いによって，図 **6.6** に示すようにピークを示した後，ひずみの増加とともに応力が低下するひずみ軟化も見られ，土の挙動を精度良く求めるためには，これらの点も考慮する必要がある。

**【例題 6.1】** 弾性体の 2 次元平面応力における応力-ひずみ関係を求めよ。

**解答** ポアソン比 $\nu$ を考慮し，縦ひずみ $\varepsilon_h$，横ひずみ $\varepsilon_v$ の関係を求めると

$$\varepsilon_h = -\nu\varepsilon_v \tag{6.7}$$

$$\varepsilon_v = -\nu\varepsilon_h \tag{6.8}$$

となる。

2 次元における横方向のひずみ $\varepsilon_x$，および縦方向のひずみ $\varepsilon_y$ は

$$\varepsilon_x = \varepsilon_h - \nu\varepsilon_v \tag{6.9}$$

$$\varepsilon_y = -\nu\varepsilon_h + \varepsilon_v \tag{6.10}$$

である。

フックの法則 $\sigma = E\varepsilon$ より，式(6.9)，(6.10)は

$$\varepsilon_x = \frac{\sigma_x}{E} - \nu\frac{\sigma_y}{E} \tag{6.11}$$

$$\varepsilon_y = -\nu\frac{\sigma_x}{E} + \frac{\sigma_y}{E} \tag{6.12}$$

となる。

また，せん断ひずみ $\gamma_{xy}$，せん断弾性係数 $G$ は，それぞれ

$$\gamma_{xy} = \frac{\tau_{xy}}{G} \tag{6.13}$$

$$G = \frac{E}{2(1+\nu)} \tag{6.14}$$

であることから

$$\gamma_{xy} = \frac{2(1+\nu)}{E}\tau_{xy} \tag{6.15}$$

である。

したがって，式(6.11)，(6.12)，(6.15)より

$$\begin{pmatrix} \varepsilon_x \\ \varepsilon_y \\ \gamma_{xy} \end{pmatrix} = \frac{1}{E}\begin{bmatrix} 1 & -\nu & 0 \\ -\nu & 1 & 0 \\ 0 & 0 & 2(1+\nu) \end{bmatrix}\begin{pmatrix} \sigma_x \\ \sigma_y \\ \tau_{xy} \end{pmatrix} \tag{6.16}$$

$$\begin{pmatrix} \varepsilon_x \\ \varepsilon_y \\ \gamma_{xy} \end{pmatrix} = \frac{1}{E} \begin{bmatrix} 1 & -\nu & 0 \\ -\nu & 1 & 0 \\ 0 & 0 & 2(1+\nu) \end{bmatrix} \begin{pmatrix} \sigma_x \\ \sigma_y \\ \tau_{xy} \end{pmatrix} \qquad (6.16)$$

Therefore, the stress-strain relationships for a two-dimensional plane-stress condition for an elastic body can be written as follows:

$$\begin{pmatrix} \sigma_x \\ \sigma_y \\ \tau_{xy} \end{pmatrix} = \frac{E}{1-\nu^2} \begin{bmatrix} 1 & \nu & 0 \\ \nu & 1 & 0 \\ 0 & 0 & \frac{1+\nu}{2} \end{bmatrix} \begin{pmatrix} \varepsilon_x \\ \varepsilon_y \\ \gamma_{xy} \end{pmatrix} \qquad (6.17)$$

---

### Charles-Augustin de Coulomb（クーロン，1736～1806）

フランスの物理学者，土木技術者。

フランス・アングレームの裕福な家に生まれる。少年時代パリの名門校 College des Quatre-Nations に学ぶ。

1761 年陸軍士官学校を卒業，カリブ海にあるマルティニーク島へ転属，ブルボン城砦の建設に 8 年間従事した。

この間，石造建築の耐久性や支持構造物の振る舞いに関する実験を行った。1776 年「建築に関するいくつかの静力学的問題に最大・最小の原理の適用に関する試み」を発表。

1874 年，針金のねじれと弾性に関する理論的研究と実験が著される。1875 年に電磁気に関する 3 報の論文を発表した。彼が発明したねじり秤を用いて，帯電した物体間に働く力を測定し，クーロンの法則を発見した。電荷の単位「クーロン」は彼の名前に由来する。

図1　クーロン

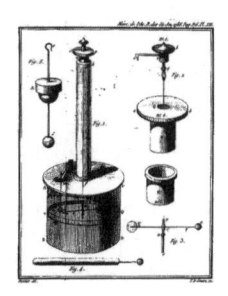

図2　ねじり秤

となり，弾性体の2次元平面応力における応力-ひずみ関係は

$$\begin{pmatrix} \sigma_x \\ \sigma_y \\ \tau_{xy} \end{pmatrix} = \frac{E}{1-\nu^2} \begin{bmatrix} 1 & \nu & 0 \\ \nu & 1 & 0 \\ 0 & 0 & \frac{1+\nu}{2} \end{bmatrix} \begin{pmatrix} \varepsilon_x \\ \varepsilon_y \\ \gamma_{xy} \end{pmatrix} \tag{6.17}$$

で表される。

## William John Macquorn Rankine（ランキン，1820〜1872）

　スコットランドの技術者，物理学者。エジンバラで生まれ，グラスゴーで死す。年少時，ランキンは体が弱かったので家で教育を受けた。1834年陸海軍士官学校で学び，その頃より数学に秀でた。叔父よりラテン語で書かれたニュートンのプリンキピアを貰った。エジンバラ大学を出た後，おそらく家が困窮していたのであろう，アイルランド鉄道の測量士の見習いになった。それが，その後ランキン法と呼ばれる技術に発達した。

　熱力学で大きな業績をあげた。長らく主流であった熱素説を否定し「エネルギー」の用語と概念を導入した。ほぼ同時期のトムソン（ケルヴィン卿），クラジウスと並んで，熱力学の基礎を作った人物だと評価されている。衝撃波伝播に関しランキン-ユゴニオの式，蒸気タービンのランキンサイクルなどが知られている。1842年に起きたベルサイユ列車事故が車軸の疲労破壊によるものだと述べた。温度の単位「蘭氏」（ランキン度）は彼の名前に因む。土質力学ではランキン土圧論で有名。

図3　ランキン　　　　図4　ベルサイユ列車事故

Rankine, W.: On the stability of loose earth. Philosophical Transactions of the Royal Society of London, **147** (1857)

# Chapter 7  Mohr's Stress Circle

## 7.1 Mohr's Stress Circle for Principal Stress

Ground deformation in the horizontal direction that arises from a local vertical (normal) stress is restrained by the surrounding soil. For this reason, it is necessary to consider the stresses in at least two-dimensions in the ground, namely, the vertical stress and the horizontal stress. When an external load acts on the ground, a shear surface is developed along line (BC) at an angle $\alpha$ with the horizontal plane as shown in **Figure 7.1** (a). If the principal stresses acting in the ground are $\sigma_v$ and $\sigma_h$, the stress acting on an element is given as in Figure 7.1 (b).

Using the equation Force=Stress×Area (assuming a unit width=1 for an element perpendicular to the plane of the paper), the vertical (normal) stress $\sigma$ and shear stress $\tau$ on the shear surface (BC in Figure 7.1) are given by Equations (7.1) and (7.2). Equations (7.1) and (7.2) balance the forces in the $\sigma$ and the $\tau$ directions.

$$\sigma \overline{BC} = \sigma_h \sin\alpha \, \overline{AB} + \sigma_v \cos\alpha \, \overline{AC} \qquad (7.1)$$
$$\tau \overline{BC} = \sigma_v \sin\alpha \, \overline{AC} - \sigma_h \cos\alpha \, \overline{AB} \qquad (7.2)$$

Dividing both sides of Equations (7.1) and (7.2) by $(\overline{BC})$ and taking $\cos\alpha = \dfrac{\overline{AC}}{\overline{BC}}$ and $\sin\alpha = \dfrac{\overline{AB}}{\overline{BC}}$, the following equations are obtained:

$$\sigma = \sigma_h \sin^2\alpha + \sigma_v \cos^2\alpha \qquad (7.3)$$
$$\tau = \sigma_v \sin\alpha \cos\alpha - \sigma_h \sin\alpha \cos\alpha \qquad (7.4)$$

Substituting $\sin 2\alpha = 2\sin\alpha \cos\alpha$, $\cos 2\alpha = \cos^2\alpha - \sin^2\alpha$ and $\sin^2\alpha$

# 7章　モールの応力円

## 7.1　主応力によるモールの応力円

　地盤内では，水平方向の変位が規制されるので，垂直応力が作用した場合，横方向にはらみ出そうとする変位を，水平方向の応力によって抑制しようとする。このため，垂直方向の応力だけではなく，水平方向の応力も考慮した，少なくとも2次元での応力を考える必要がある。

　地盤に外力が作用し，地盤内に水平面と $\alpha$ の角度の bc（BC）面に沿ってずれ（せん断面）が生じたとする。このとき，地盤内に主応力として $\sigma_v$，$\sigma_h$ が作用すると考えると，このせん断面を含む要素 ABDC に作用する応力は図7.1 となる。

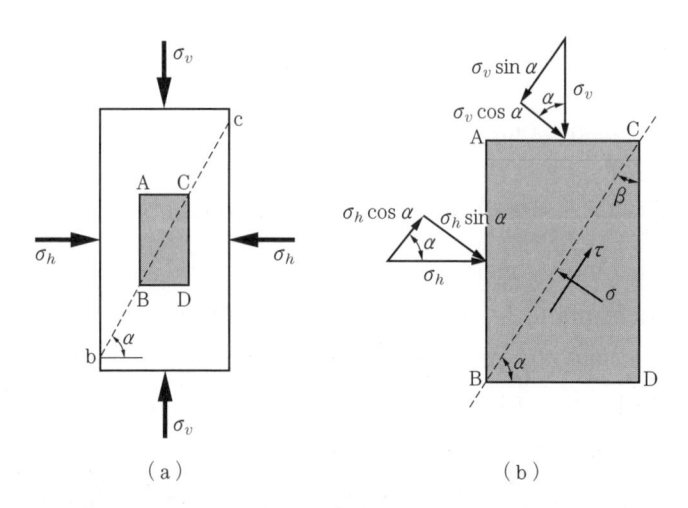

**Figure 7.1**　Principal stresses acting on an infinitesimal element
　図7.1　微小要素に作用する主応力

$+\cos^2\alpha=1$ into Equations (7.3) and (7.4) yields:

$$\sigma=\frac{\sigma_h}{2}(1+\cos 2\alpha)+\frac{\sigma_v}{2}(1-\cos 2\alpha)=\frac{\sigma_v+\sigma_h}{2}+\frac{\sigma_v-\sigma_h}{2}\cos 2\alpha$$

(7.5)

$$\tau=\left(\frac{\sigma_v-\sigma_h}{2}\right)\sin 2\alpha \qquad (7.6)$$

Substituting Equations (7.5) and (7.6) gives

$$\left(\sigma-\frac{\sigma_v+\sigma_h}{2}\right)^2+\tau^2$$

and leads to Equation (7.7):

$$\left(\sigma-\frac{\sigma_v+\sigma_h}{2}\right)^2+\tau^2=\left(\frac{\sigma_v+\sigma_h}{2}+\frac{\sigma_v-\sigma_h}{2}\cos 2\alpha-\frac{\sigma_v+\sigma_h}{2}\right)^2$$

$$+\left\{\left(\frac{\sigma_v-\sigma_h}{2}\right)\sin 2\alpha\right\}^2$$

$$=\left\{\left(\frac{\sigma_v-\sigma_h}{2}\right)\cos 2\alpha\right\}^2+\left\{\left(\frac{\sigma_v-\sigma_h}{2}\right)\sin 2\alpha\right\}^2$$

$$=\left(\frac{\sigma_v-\sigma_h}{2}\right)^2(\cos^2 2\alpha+\sin^2 2\alpha)$$

$$=\left(\frac{\sigma_v-\sigma_h}{2}\right)^2 \qquad (7.7)$$

The term $\left(\sigma-\frac{\sigma_v+\sigma_h}{2}\right)^2+\tau^2=\left(\frac{\sigma_v-\sigma_h}{2}\right)^2$ of Equation (7.7) is the equation of a circle having its center at $\left(\frac{\sigma_x+\sigma_y}{2}, 0\right)$ and a radius of $\left(\frac{\sigma_x-\sigma_y}{2}\right)$. The circle drawn by this Equation (7.7) is called Mohr's stress circle. Now, consider a Mohr's stress circle as shown in **Figure 7.2**. The coordinate $(\sigma, \tau)$ of the intersection point between Mohr's stress circle and the line, which makes an angle $\alpha$ with horizontal plane from point A of $\sigma_h$, is given by the following equations:

$$\overline{AE}=\overline{CE}=\frac{\sigma_v-\sigma_h}{2} \qquad (7.8)$$

$$\tau=\overline{CE}\sin 2\beta \qquad (7.9)$$

ここで，せん断面 bc（BC）に発生する垂直応力 $\sigma$，せん断応力 $\tau$ を求めてみる。

要素の奥行きを単位幅（=1）として，垂直応力 $\sigma$ 方向およびせん断応力 $\tau$ 方向の力の釣合いを求めると，力＝応力×面積（ここで，幅は単位幅 1 を考える）より，

$\sigma$ 方向の力の釣合い

$$\sigma \overline{BC} = \sigma_h \sin\alpha \, \overline{AB} + \sigma_v \cos\alpha \, \overline{AC} \tag{7.1}$$

$\tau$ 方向の力の釣合い

$$\tau \overline{BC} = \sigma_v \sin\alpha \, \overline{AC} - \sigma_h \cos\alpha \, \overline{AB} \tag{7.2}$$

である。

ここで，$\cos\alpha = \dfrac{\overline{AC}}{\overline{BC}}$，$\sin\alpha = \dfrac{\overline{AB}}{\overline{BC}}$ を考慮し，式(7.1)，(7.2)の両辺を $\overline{BC}$ で除すと

$$\sigma = \sigma_h \sin^2\alpha + \sigma_v \cos^2\alpha \tag{7.3}$$

$$\tau = \sigma_v \sin\alpha \cos\alpha - \sigma_h \sin\alpha \cos\alpha \tag{7.4}$$

となる。

式(7.3)，(7.4)に $\sin 2\alpha = 2\sin\alpha \cos\alpha$，$\cos 2\alpha = \cos^2\alpha - \sin^2\alpha$，$\sin^2\alpha + \cos^2\alpha = 1$ の関係を用いて変形すると

$$\sigma = \frac{\sigma_h}{2}(1 + \cos 2\alpha) + \frac{\sigma_v}{2}(1 - \cos 2\alpha) = \frac{\sigma_v + \sigma_h}{2} + \frac{\sigma_v - \sigma_h}{2} \cos 2\alpha \tag{7.5}$$

$$\tau = \left(\frac{\sigma_v - \sigma_h}{2}\right) \sin 2\alpha \tag{7.6}$$

が求まる。

ここで，$\left(\sigma - \dfrac{\sigma_v + \sigma_h}{2}\right)^2 + \tau^2$ を考慮し，この式に式(7.5)，(7.6)を代入すると

$$\left(\sigma - \frac{\sigma_v + \sigma_h}{2}\right)^2 + \tau^2 = \left(\frac{\sigma_v + \sigma_h}{2} + \frac{\sigma_v - \sigma_h}{2} \cos 2\alpha - \frac{\sigma_v + \sigma_h}{2}\right)^2$$
$$+ \left\{\left(\frac{\sigma_v - \sigma_h}{2}\right) \sin 2\alpha\right\}^2$$

## Chapter 7  Mohr's Stress Circle

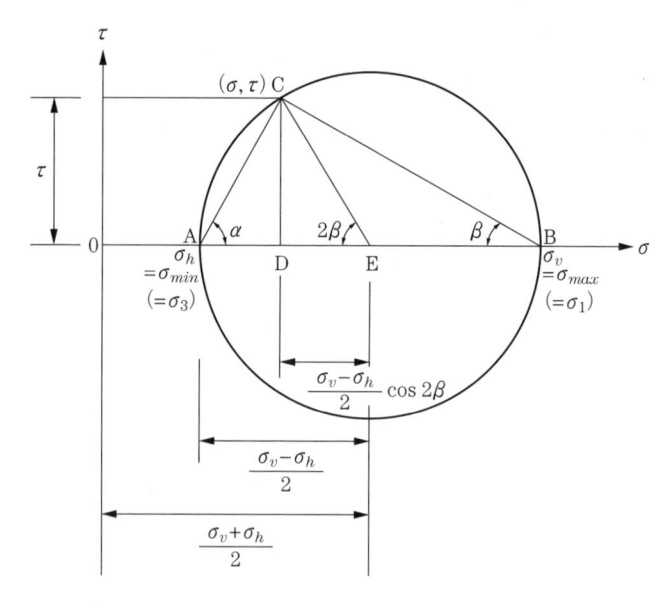

**Figure 7.2**  Mohr's stress circle of principal stresses
図 7.2  主応力でのモールの応力円

$$\sigma = \frac{\tau}{\tan \alpha} + \sigma_h \tag{7.10}$$

Here, triangle △ABC in Figure 7.1 and triangle △ABC in Figure 7.2 are similar. If vertical and horizontal stresses act in a ground such that $\sigma_v > \sigma_h$, then the stresses ($\sigma$, $\tau$) on a surface with angle $\alpha$ can be calculated by using Mohr's stress circle. In this case, the value of the horizontal stress of point (A) is $\sigma_h$. A straight line is drawn from point (A) with angle $\alpha$ to a point (C) on the circle. The horizontal stress is the abscissa coordinate of point (C) which is the intersection of this straight line and Mohr's stress circle.

The larger principal stress is called the maximum principal stress $\sigma_{max}$ and the smaller principal stress is called the minimum principal stress $\sigma_{min}$. In soil mechanics, the maximum principal stress is represented by $\sigma_1$ and the minimum principal stress is represented

$$= \left\{\left(\frac{\sigma_v - \sigma_h}{2}\right)\cos 2\alpha\right\}^2 + \left\{\left(\frac{\sigma_v - \sigma_h}{2}\right)\sin 2\alpha\right\}^2$$

$$= \left(\frac{\sigma_v - \sigma_h}{2}\right)^2 (\cos^2 2\alpha + \sin^2 2\alpha)$$

$$= \left(\frac{\sigma_v - \sigma_h}{2}\right)^2 \tag{7.7}$$

となる。

式(7.7)が示す $\left(\sigma - \frac{\sigma_v + \sigma_h}{2}\right)^2 + \tau^2 = \left(\frac{\sigma_v - \sigma_h}{2}\right)^2$ は，中心 $\left(\frac{\sigma_x + \sigma_y}{2}, 0\right)$，半径 $\left(\frac{\sigma_x - \sigma_y}{2}\right)$ の円の方程式を示し，この式(7.7)から求まる円をモールの応力円という。

図7.2に示すモールの応力円を考え，$\sigma_h$ のA点の位置から水平面に対し角度 $\alpha$ 傾いた直線とモールの応力円との交点の座標 $(\sigma, \tau)$ は

$$\overline{\text{AE}} = \overline{\text{CE}} = \frac{\sigma_v - \sigma_h}{2} \tag{7.8}$$

$$\tau = \overline{\text{CE}} \sin 2\beta \tag{7.9}$$

$$\sigma = \frac{\tau}{\tan \alpha} + \sigma_h \tag{7.10}$$

で表される。

ここで，図7.1の△ABCと図7.2の△ABCとは相似であり，垂直応力 $\sigma_v$，水平応力 $\sigma_h$ が作用する地盤内（$\sigma_v > \sigma_h$）において，角度 $\alpha$ 傾いた面に作用する応力 $(\sigma, \tau)$ をモールの応力円を用いて求めるには，$\sigma_h$ を示すA点から角度 $\alpha$ の直線を引き，この直線とモールの応力円との交点Cの座標から求めればよいことになる。

なお，主応力は値が大きい方を最大主応力 $\sigma_{max}$，小さい方を最小主応力 $\sigma_{min}$ といい，土質力学では最大主応力を $\sigma_1$，最小主応力を $\sigma_3$ で表す。

【例題 7.1】 図7.3に示す，一辺5 cm で長さ15 cm の直方体の周囲全体に $P = 7.5$ kN の荷重を作用させたとき，水平面と60°の角度でせん断面が生じた。せん断面で生じる垂直応力 $\sigma$ とせん断応力 $\tau$ を求めよ。なお，直方体の

by $\sigma_3$.

【**Example 7.1**】 A load $P=7.5$ kN is applied to a rectangular body of length 15 cm and width 5 cm as shown in **Figure 7.3**. Failure occurs at an angle of 60° with the horizontal. No shear stress is applied to the surface of the rectangular body. Calculate the normal stress and shear stress on the failure surface (dashed line in Figure 7.3).

**Solution** As there is no shear stress on any of the surfaces of the rectangular body, the normal stresses on each surface are the principal stresses. The principal stress on each surface is calculated as follows:
The normal stress in vertical direction is

$$\sigma_v = \frac{P}{A} = \frac{7.5}{0.05 \times 0.05} = 3\,000 \text{ kN/m}^2 \tag{7.11}$$

The normal stress in horizontal direction is

$$\sigma_h = \frac{P}{A} = \frac{7.5}{0.05 \times 0.15} = 1\,000 \text{ kN/m}^2 \tag{7.12}$$

By comparing the magnitude of the normal stress components, the maximum principal stress $\sigma_1$ and minimum principal stress $\sigma_3$ are

$\sigma_1 = 3\,000$ kN/m²
$\sigma_3 = 1\,000$ kN/m²

---

**Jean Louis Marie Poiseuille**(ポアズイユ, 1799〜1869)

フランスの物理学者,生理学者。1815〜1816年パリのエコール・ポリテクニークで物理と数学を学んだ。円筒管内部を流れる非圧縮性の粘性流体の層流流れについて1838年に実験し,1840年頃にポアズイユの定理を数式化した(ゴットヒルフ・ハーゲン Gotthilf H. L. Hagen(1797〜1884)も独立して研究したので,ハーゲン＝ポアズイユの式とも呼ばれる)。CGS単位系の粘度の単位ポアズは,ポアズイユの名に因んでいる。

図5 ポアズイユ

## 7.1 Mohr's Stress Circle for Principal Stress

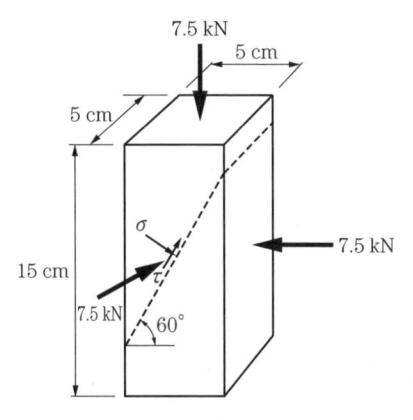

**Figure 7.3** Stresses acting on a shear surface in a rectangular body
図7.3 せん断面に作用する応力

各面においてせん断応力の発生はないとする。

**解答** 直方体各面でせん断応力の発生がないので，各面で発生する垂直応力は主応力と考えられる。各面の主応力を求めると，
垂直方向は

$$\sigma_v = \frac{P}{A} = \frac{7.5}{0.05 \times 0.05} = 3\,000 \text{ kN/m}^2 \tag{7.11}$$

水平方向は

$$\sigma_h = \frac{P}{A} = \frac{7.5}{0.05 \times 0.15} = 1\,000 \text{ kN/m}^2 \tag{7.12}$$

となる。

垂直応力の大小を考え，最大主応力 $\sigma_1$，最小主応力 $\sigma_3$ を求めると

$\sigma_1 = 3\,000 \text{ kN/m}^2$

$\sigma_3 = 1\,000 \text{ kN/m}^2$

である。

図7.4に示す主応力によるモールの応力円より，水平面と60°の面で発生する $(\sigma, \tau)$ は，モール円上のC点の座標を求めればよいことになる。

$\overline{\text{AE}} = \overline{\text{CE}} = \dfrac{\sigma_v - \sigma_h}{2} = 1\,000 \text{ kN/m}^2$

$\alpha = 60°$

$2\beta = 60°$

## 22 Chapter 7 Mohr's Stress Circle

Next, the principal stresses and Mohr's stress circle are used to find $\sigma$ and $\tau$ that act on a plane that makes an angle of 60° with the horizontal. **Figure 7.4** shows that is done by calculating the coordinate of point (C) on the Mohr's circle.

From

$$\overline{AE} = \overline{CE} = \frac{\sigma_v - \sigma_h}{2} = 1\,000 \text{ kN/m}^2$$

$$\alpha = 60°$$

and

$$2\beta = 60°$$

gives

$$\tau = \overline{CE} \sin 2\beta = 866 \text{ kN/m}^2$$

$$\sigma = \frac{\tau}{\tan\alpha} + \sigma_h = 1\,500 \text{ kN/m}^2$$

## 7.2 Conjugate Relationships of Shear Stresses

Now we consider normal and shear stresses in two-dimensions for a ground subjected to an external force. To find the relationships between $\tau_{xy}$ and $\tau_{yx}$, we examine the four stresses ($\sigma_x$, $\sigma_y$, $\tau_{xy}$, $\tau_{yx}$) that act on an infinitesimal element in the ground as shown in **Figure 7.5**. Stresses at a point of an infinitesimal distance $dx$ and $dy$ are given by

$$\sigma_x + \frac{\partial \sigma_x}{\partial x} dx, \quad \sigma_y + \frac{\partial \sigma_y}{\partial y} dy, \quad \tau_{xy} + \frac{\partial \tau_{xy}}{\partial x} dx, \quad \tau_{yx} + \frac{\partial \tau_{yx}}{\partial y} dy$$

Moments of $\sigma_x$ and $\sigma_y$ are to zero as they pass through the center of gravity. Taking the equilibrium of moment about the center of gravity (G) of the infinitesimal element, yields

$$\left(\tau_{xy} + \frac{\partial \tau_{xy}}{\partial x} dx\right) dy \frac{dx}{2} + \tau_{xy} dy \frac{dx}{2} = \left(\tau_{yx} + \frac{\partial \tau_{yx}}{\partial y} dy\right) dx \frac{dy}{2}$$

$$+ \tau_{yx} dx \frac{dy}{2} \quad (7.13)$$

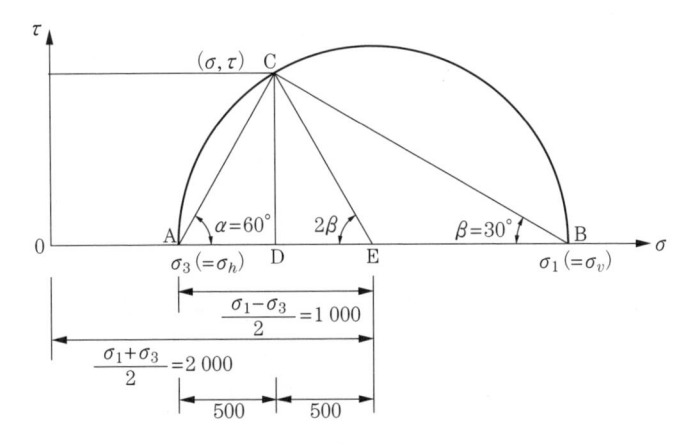

**Figure 7.4** Mohr's stress circle for Example 7.1
図 7.4 モールの応力円

より

$$\tau = \overline{\text{CE}} \sin 2\beta = 866 \text{ kN/m}^2$$

$$\sigma = \frac{\tau}{\tan \alpha} + \sigma_h = 1\,500 \text{ kN/m}^2$$

となる。

## 7.2 せん断応力の共役関係

2次元問題において，外力が作用し地盤内に垂直応力とせん断応力が作用する場合を考えてみる。このとき，地盤内の微小要素には，図 7.5 に示すような $\sigma_x$, $\sigma_y$, $\tau_{xy}$, $\tau_{yx}$ の四つの応力が作用する。ここで，$\tau_{xy}$ と $\tau_{yx}$ の関係について考えてみる。

微小距離 $dx$, $dy$ だけ離れた地点における各応力は

$$\sigma_x + \frac{\partial \sigma_x}{\partial x} dx, \quad \sigma_y + \frac{\partial \sigma_y}{\partial y} dy, \quad \tau_{xy} + \frac{\partial \tau_{xy}}{\partial x} dx, \quad \tau_{yx} + \frac{\partial \tau_{yx}}{\partial y} dy$$

である。

ここで，微小要素の重心 G についてモーメントの釣合いを考えてみると，

## Chapter 7  Mohr's Stress Circle

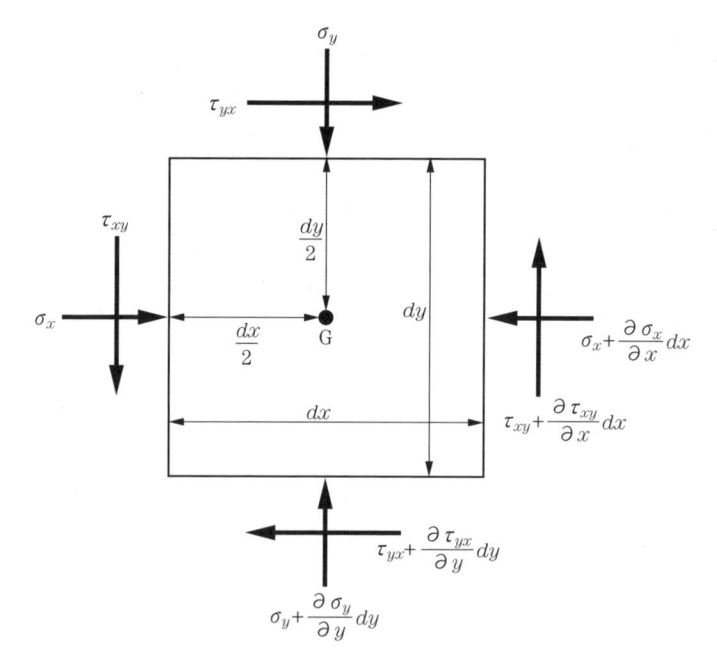

**Figure 7.5**  Stresses acting on an infinitesimal element
図 7.5  微小要素に作用する応力

$$\left(\frac{\partial \tau_{xy}}{\partial x}\right)dy\frac{dx^2}{2} + \tau_{xy}\,dxdy = \left(\frac{\partial \tau_{yx}}{\partial y}\right)dx\frac{dy^2}{2} + \tau_{yx}\,dxdy \quad (7.14)$$

Setting the squared terms to zero (i.e., $dx^2 = dy^2 \approx 0$), the following relationship is obtained

$$\tau_{xy} = \tau_{yx} \quad (7.15)$$

Equation (7.15) is the conjugate relationship of shear stress.

## 7.3  Mohr's Stress Circle Considering Shear Stresses

Now we consider both shear stresses $\tau_{xy}$ and $\tau_{yx}$ and the normal stresses (vertical component $\sigma_y$ and horizontal component $\sigma_x$) acting on a ground. For shear stresses in a ground, the conjugate relationship

$\sigma_x$, $\sigma_y$ は重心上にありモーメントは 0 となるため

$$\left(\tau_{xy}+\frac{\partial \tau_{xy}}{\partial x}\,dx\right)dy\,\frac{dx}{2}+\tau_{xy}\,dy\,\frac{dx}{2}=\left(\tau_{yx}+\frac{\partial \tau_{yx}}{\partial y}\,dy\right)dx\,\frac{dy}{2}$$
$$+\tau_{yx}\,dx\,\frac{dy}{2} \qquad (7.13)$$

$$\left(\frac{\partial \tau_{xy}}{\partial x}\right)dy\,\frac{dx^2}{2}+\tau_{xy}\,dxdy=\left(\frac{\partial \tau_{yx}}{\partial y}\right)dx\,\frac{dy^2}{2}+\tau_{yx}\,dxdy \qquad (7.14)$$

である。

式(7.14)において 2 次の微小項を 0 ($dx^2=dy^2\approx 0$) と置くと

$$\tau_{xy}=\tau_{yx} \qquad (7.15)$$

の関係が求まる。

これをせん断応力の共役関係という。

## 7.3　せん断応力を考慮したモールの応力円

2 次元問題において地盤内に垂直方向の応力 $\sigma_y$，水平方向の応力 $\sigma_x$ と併せてせん断応力 $\tau_{xy}$，$\tau_{yx}$ が作用する場合について考えてみる。地盤内のせん断応力を考える場合，せん断応力の共役関係より $\tau_{xy}=\tau_{yx}$ と考えることができるため，外力の作用によって地盤内に作用する応力は，図 7.6 に示すものとなる。このとき，ある角度 $\alpha$ 傾いた bc（BC）面に発生する応力 $\sigma$，$\tau$ を求めてみる。

ここで，$\sigma$ 方向，$\tau$ 方向の力の釣合いを求めると，
$\sigma$ 方向の力の釣合いは

$$\sigma\,\overline{BC}=\sigma_x\sin\alpha\,\overline{AB}+\sigma_y\cos\alpha\,\overline{AC}+\tau_{xy}\sin\alpha\,\overline{AC}+\tau_{xy}\cos\alpha\,\overline{AB}$$
$$(7.16)$$

$\tau$ 方向の力の釣合いは

$$\tau\,\overline{BC}=\sigma_y\sin\alpha\,\overline{AC}-\sigma_x\cos\alpha\,\overline{AB}+\tau_{xy}\sin\alpha\,\overline{AB}-\tau_{xy}\cos\alpha\,\overline{AC}$$
$$(7.17)$$

である。

($\tau_{xy} = \tau_{yx}$) holds. The stresses acting on a ground caused by an external force are as shown in **Figure 7.6**. Let's calculate the stresses ($\sigma$, $\tau$) acting on the surface (BC) in Figure 7.6 oriented with an angle $\alpha$. Balancing the forces in the $\sigma$ direction yields:

$$\sigma \overline{BC} = \sigma_x \sin\alpha \, \overline{AB} + \sigma_y \cos\alpha \, \overline{AC} + \tau_{xy} \sin\alpha \, \overline{AC} + \tau_{xy} \cos\alpha \, \overline{AB} \tag{7.16}$$

Balancing the forces in the $\tau$ direction gives:

$$\tau \overline{BC} = \sigma_y \sin\alpha \, \overline{AC} - \sigma_x \cos\alpha \, \overline{AB} + \tau_{xy} \sin\alpha \, \overline{AB} - \tau_{xy} \cos\alpha \, \overline{AC} \tag{7.17}$$

Taking $\cos\alpha = \dfrac{\overline{AC}}{\overline{BC}}$ and $\sin\alpha = \dfrac{\overline{AB}}{\overline{BC}}$, and dividing both sides of Equations (7.16) and (7.17) by $\overline{BC}$, we obtain

$$\sigma = \sigma_x \sin^2\alpha + \sigma_y \cos^2\alpha + 2\tau_{xy} \sin\alpha \cos\alpha \tag{7.18}$$

$$\tau = \sigma_y \sin\alpha \cos\alpha - \sigma_x \sin\alpha \cos\alpha + \tau_{xy} \sin^2\alpha - \tau_{xy} \cos^2\alpha \tag{7.19}$$

Substituting $\sin 2\alpha = 2\sin\alpha \cos\alpha$, $\cos 2\alpha = \cos^2\alpha - \sin^2\alpha$ and $\sin^2\alpha + \cos^2\alpha = 1$ into Equations (7.18) and (7.19), $\sigma$ and $\tau$ are

$$\sigma = \frac{\sigma_x}{2}(1 + \cos 2\alpha) + \frac{\sigma_y}{2}(1 - \cos 2\alpha) + \tau_{xy} \sin 2\alpha$$

$$= \frac{\sigma_x + \sigma_y}{2} + \frac{\sigma_x - \sigma_y}{2} \cos 2\alpha + \tau_{xy} \sin 2\alpha \tag{7.20}$$

$$\tau = -(\sigma_x - \sigma_y) \sin\alpha \cos\alpha - \tau_{xy}(\cos^2\alpha - \sin^2\alpha)$$

$$= -\frac{\sigma_x - \sigma_y}{2} \sin 2\alpha - \tau_{xy} \cos 2\alpha \tag{7.21}$$

Substituting Equations (7.20) and (7.21) into $\left(\sigma - \dfrac{\sigma_x + \sigma_y}{2}\right)^2 + \tau^2$, yields

$$\left(\sigma - \frac{\sigma_x + \sigma_y}{2}\right)^2 + \tau^2 = \left(\frac{\sigma_x + \sigma_y}{2} + \frac{\sigma_x - \sigma_y}{2} \cos 2\alpha\right.$$
$$\left. + \tau_{xy} \sin 2\alpha - \frac{\sigma_x + \sigma_y}{2}\right)^2$$
$$+ \left(\frac{\sigma_x - \sigma_y}{2} \sin 2\alpha + \tau_{xy} \cos 2\alpha\right)^2$$

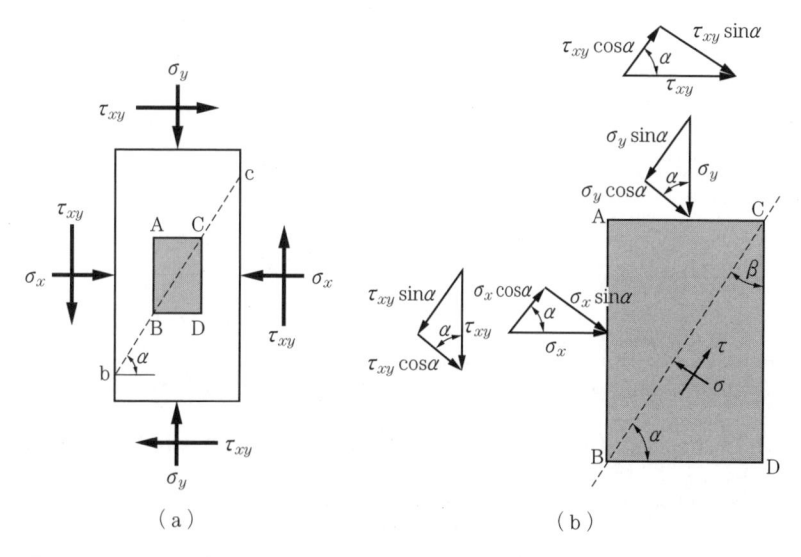

## 7.3 Mohr's Stress Circle Considering Shear Stresses

**Figure 7.6** Shear stress and normal stress acting on an infinitesimal element
図 7.6 微小要素に作用するせん断応力と垂直応力

ここで，$\cos\alpha = \dfrac{\overline{AC}}{\overline{BC}}$, $\sin\alpha = \dfrac{\overline{AB}}{\overline{BC}}$ を考慮し，式(7.16)，(7.17)の両辺を $\overline{BC}$ で除すと

$$\sigma = \sigma_x \sin^2\alpha + \sigma_y \cos^2\alpha + 2\tau_{xy}\sin\alpha\cos\alpha \tag{7.18}$$

$$\tau = \sigma_y \sin\alpha\cos\alpha - \sigma_x \sin\alpha\cos\alpha + \tau_{xy}\sin^2\alpha - \tau_{xy}\cos^2\alpha \tag{7.19}$$

となる。

式(7.18)，(7.19)に $\sin 2\alpha = 2\sin\alpha\cos\alpha$, $\cos 2\alpha = \cos^2\alpha - \sin^2\alpha$, $\sin^2\alpha + \cos^2\alpha = 1$ の関係を用いて変形すると

$$\begin{aligned}\sigma &= \frac{\sigma_x}{2}(1+\cos 2\alpha) + \frac{\sigma_y}{2}(1-\cos 2\alpha) + \tau_{xy}\sin 2\alpha \\ &= \frac{\sigma_x+\sigma_y}{2} + \frac{\sigma_x-\sigma_y}{2}\cos 2\alpha + \tau_{xy}\sin 2\alpha \end{aligned} \tag{7.20}$$

$$\begin{aligned}\tau &= -(\sigma_x - \sigma_y)\sin\alpha\cos\alpha - \tau_{xy}(\cos^2\alpha - \sin^2\alpha) \\ &= -\frac{\sigma_x - \sigma_y}{2}\sin 2\alpha - \tau_{xy}\cos 2\alpha \end{aligned} \tag{7.21}$$

28　Chapter 7　Mohr's Stress Circle

$$= \left(\frac{\sigma_x - \sigma_y}{2} \cos 2\alpha + \tau_{xy} \sin 2\alpha\right)^2$$

$$+ \left(\frac{\sigma_x - \sigma_y}{2} \sin 2\alpha + \tau_{xy} \cos 2\alpha\right)^2$$

$$= \left(\frac{\sigma_x - \sigma_y}{2}\right)^2 (\cos^2 2\alpha + \sin^2 2\alpha)$$

$$+ \tau_{xy}^2 (\sin^2 2\alpha + \cos^2 2\alpha)$$

$$= \left(\frac{\sigma_x - \sigma_y}{2}\right)^2 + \tau_{xy}^2 \qquad (7.22)$$

The expression $\left(\sigma - \frac{\sigma_x + \sigma_y}{2}\right)^2 + \tau^2 = \left(\frac{\sigma_x - \sigma_y}{2}\right)^2 + \tau_{xy}^2$ from Equation (7.22) is the equation for a Mohr's stress circle with the center at $\left(\frac{\sigma_x + \sigma_y}{2}, 0\right)$ and a radius of $\sqrt{\left(\frac{\sigma_x - \sigma_y}{2}\right)^2 + \tau_{xy}^2}$.

By using the Mohr's stress circle given by Equation (7.22), the maximum principal stress $\sigma_{max}$ ($=\sigma_1$), the minimum principal stress $\sigma_{mix}$ ($=\sigma_3$) and the maximum (minimum) principal shear stresses $\tau_{max}$ ($\tau_{mix}$) can be determined (**Figure 7.7**).

The $\sigma_{max}$ and $\sigma_{min}$ are the stresses at point (B) and (A), respectively, for which the shear stresses are zero.

Because, $\overline{OE} = \frac{\sigma_x + \sigma_y}{2}$, $\overline{DE} = \frac{\sigma_x - \sigma_y}{2}$, $\overline{EB} = \sqrt{\left(\frac{\sigma_x - \sigma_y}{2}\right)^2 + \tau_{xy}^2}$,

from $\overline{OB} = \overline{OE} + \overline{EB}$, the maximum principal stress $\sigma_{max}$ is given by:

$$\sigma_{max}(=\sigma_1) = \frac{\sigma_x + \sigma_y}{2} + \sqrt{\left(\frac{\sigma_x - \sigma_y}{2}\right)^2 + \tau_{xy}^2} \qquad (7.23)$$

and from $\overline{OA} = \overline{OE} - \overline{EB}$, the minimum principal stress $\sigma_{min}$ is given by:

$$\sigma_{min}(=\sigma_3) = \frac{\sigma_x + \sigma_y}{2} - \sqrt{\left(\frac{\sigma_x - \sigma_y}{2}\right)^2 + \tau_{xy}^2} \qquad (7.24)$$

$\tau_{max}$ and $\tau_{min}$ are the stresses at points (C) and (D), respectively. By using the radius of Mohr's stress circle, $\tau_{max}$ and $\tau_{min}$ are

## 7.3 Mohr's Stress Circle Considering Shear Stresses

が求まる。

ここで，$\left(\sigma-\dfrac{\sigma_x+\sigma_y}{2}\right)^2+\tau^2$ を考え，この式に式(7.20)，(7.21)を代入すると

$$
\begin{aligned}
\left(\sigma-\frac{\sigma_x+\sigma_y}{2}\right)^2+\tau^2 &= \left(\frac{\sigma_x+\sigma_y}{2}+\frac{\sigma_x-\sigma_y}{2}\cos 2\alpha \right. \\
&\quad \left. +\tau_{xy}\sin 2\alpha-\frac{\sigma_x+\sigma_y}{2}\right)^2 \\
&\quad +\left(\frac{\sigma_x-\sigma_y}{2}\sin 2\alpha+\tau_{xy}\cos 2\alpha\right)^2 \\
&= \left(\frac{\sigma_x-\sigma_y}{2}\cos 2\alpha+\tau_{xy}\sin 2\alpha\right)^2 \\
&\quad +\left(\frac{\sigma_x-\sigma_y}{2}\sin 2\alpha+\tau_{xy}\cos 2\alpha\right)^2 \\
&= \left(\frac{\sigma_x-\sigma_y}{2}\right)^2(\cos^2 2\alpha+\sin^2 2\alpha) \\
&\quad +\tau_{xy}^{\,2}(\sin^2 2\alpha+\cos^2 2\alpha) \\
&= \left(\frac{\sigma_x-\sigma_y}{2}\right)^2+\tau_{xy}^{\,2} \quad\quad (7.22)
\end{aligned}
$$

となる。

式(7.22)が示す $\left(\sigma-\dfrac{\sigma_x+\sigma_y}{2}\right)^2+\tau^2=\left(\dfrac{\sigma_x-\sigma_y}{2}\right)^2+\tau_{xy}^{\,2}$ は，中心 $\left(\dfrac{\sigma_x+\sigma_y}{2},\, 0\right)$，半径 $\sqrt{\left(\dfrac{\sigma_x-\sigma_y}{2}\right)^2+\tau_{xy}^{\,2}}$ のモールの応力円の方程式を示す。

この式(7.22)から求まるモールの応力円を基に，最大主応力 $\sigma_{max}$ ($=\sigma_1$)，最小主応力 $\sigma_{min}$ ($=\sigma_3$) および最大（最小）主せん断応力 $\tau_{max}$ ($\tau_{min}$) を求めてみる（図 **7.7**）。

$\sigma_{max}$, $\sigma_{min}$ は，せん断応力が 0 の A 点および B 点である。

ここで，$\overline{\text{OE}}=\dfrac{\sigma_x+\sigma_y}{2}$，$\overline{\text{DE}}=\dfrac{\sigma_x-\sigma_y}{2}$，$\overline{\text{EB}}=\sqrt{\left(\dfrac{\sigma_x-\sigma_y}{2}\right)^2+\tau_{xy}^{\,2}}$ であるので，最大主応力 $\sigma_{max}$ は，$\overline{\text{OB}}=\overline{\text{OE}}+\overline{\text{EB}}$ より

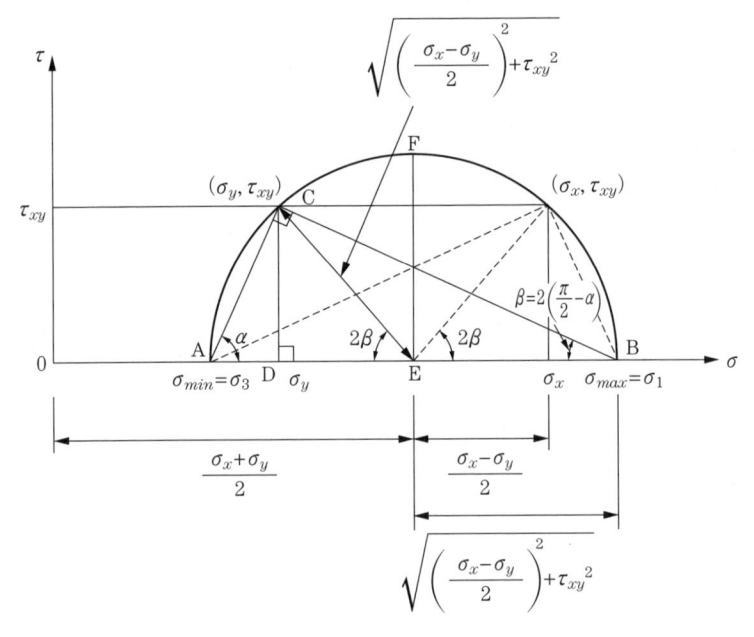

**Figure 7.7** Mohr's stress circle considering shear and normal stresses
図7.7 せん断応力と垂直応力を考慮したモールの応力円

$$\binom{\tau_{max}}{\tau_{min}} = \pm\sqrt{\left(\frac{\sigma_x-\sigma_y}{2}\right)^2+\tau_{xy}^2} \qquad (7.25)$$

The angles of the minimum principal stress, maximum principal stress and principal shear stress can be calculated from $\sigma_{min}$ at point (A), $\sigma_{max}$ at point (B) and from $\tau_{max}$ at point (C), which is 45°. The angle ($2\beta$), at the center of the circle, can be calculated from Figure 7.7 as follows:

$$\tan 2\beta = \frac{\tau_{xy}}{\frac{\sigma_x-\sigma_y}{2}} \qquad (7.26)$$

Therefore, the angles $\alpha$ and $\beta$ of the principal surfaces are

$$\alpha = \frac{\pi}{2} - \frac{1}{2}\tan^{-1}\frac{2\tau_{xy}}{\sigma_x-\sigma_y} \qquad (7.27)$$

$$\sigma_{max}(=\sigma_1)=\frac{\sigma_x+\sigma_y}{2}+\sqrt{\left(\frac{\sigma_x-\sigma_y}{2}\right)^2+\tau_{xy}^2} \tag{7.23}$$

最小主応力 $\sigma_{min}$ は，$\overline{OA}=\overline{OE}-\overline{EB}$ より

$$\sigma_{min}(=\sigma_3)=\frac{\sigma_x+\sigma_y}{2}-\sqrt{\left(\frac{\sigma_x-\sigma_y}{2}\right)^2+\tau_{xy}^2} \tag{7.24}$$

となる。

また，$\tau_{max}$，$\tau_{min}$ は，それぞれ C，D 点であるので，モールの応力円の半径より

$$\begin{pmatrix}\tau_{max}\\ \tau_{min}\end{pmatrix}=\pm\sqrt{\left(\frac{\sigma_x-\sigma_y}{2}\right)^2+\tau_{xy}^2} \tag{7.25}$$

となる。

最小主応力および最大主応力と主せん断応力とがなす角度は，$\sigma_{min}$ の A 点および $\sigma_{min}$ の B 点と，$\tau_{max}$ の C 点とのなす角度から求まり，45° である。

また，中心角 $2\beta$ は図 7.7 を考慮し

$$\tan 2\beta = \frac{\tau_{xy}}{\dfrac{\sigma_x-\sigma_y}{2}} \tag{7.26}$$

で示されることより，主応力面のなす角度 $\alpha$，$\beta$ は

$$\alpha=\frac{\pi}{2}-\frac{1}{2}\tan^{-1}\frac{2\tau_{xy}}{\sigma_x-\sigma_y} \tag{7.27}$$

$$\beta=\frac{1}{2}\tan^{-1}\frac{2\tau_{xy}}{\sigma_x-\sigma_y} \tag{7.28}$$

となる。

【例題 7.2】 $\sigma_x=500\,\mathrm{kN/m^2}$，$\sigma_y=100\,\mathrm{kN/m^2}$，$\tau_{xy}=200\,\mathrm{kN/m^2}$ のとき，主応力とその方向を求めよ。

**解答** 図 7.8 にモールの応力円を示す。このモールの応力円より，最大主応力 $\sigma_1$，最小主応力 $\sigma_3$ は

$$\begin{pmatrix}\sigma_1\\ \sigma_3\end{pmatrix}=\frac{\sigma_x+\sigma_y}{2}\pm\sqrt{\left(\frac{\sigma_x-\sigma_y}{2}\right)^2+\tau_{xy}^2} \tag{7.29}$$

$$\beta = \frac{1}{2}\tan^{-1}\frac{2\tau_{xy}}{\sigma_x - \sigma_y} \tag{7.28}$$

【**Example 7.2**】 Calculate the principal stresses and their direction when $\sigma_x = 500 \text{ kN/m}^2$, $\sigma_y = 100 \text{ kN/m}^2$ and $\tau_{xy} = 200 \text{ kN/m}^2$.

**Solution** The Mohr's stress circle for this case is shown in **Figure 7.8**. From this Mohr's stress circle, the maximum principal stress $\sigma_1$ and minimum principal stress $\sigma_3$ can be calculated from

$$\begin{pmatrix}\sigma_1\\ \sigma_3\end{pmatrix} = \frac{\sigma_x + \sigma_y}{2} \pm \sqrt{\left(\frac{\sigma_x - \sigma_y}{2}\right)^2 + \tau_{xy}^2} \tag{7.29}$$

$$= \frac{500 + 100}{2} \pm \sqrt{\left(\frac{500 - 100}{2}\right)^2 + 200^2} = 300 \pm 282.84 \text{ kN/m}^2 \tag{7.30}$$

Therefore, the maximum and minimum principal stresses are
$\sigma_1 = 582.84 \text{ kN/m}^2$
$\sigma_3 = 17.16 \text{ kN/m}^2$

The orientation of the principal surface is given by the following angles

$$\beta = \frac{1}{2}\tan^{-1}\frac{2\tau_{xy}}{\sigma_x - \sigma_y} = \frac{1}{2}\tan^{-1}\frac{2 \times 200}{400} = 22.5° \tag{7.31}$$

$$\alpha = \frac{\pi}{2} - \frac{1}{2}\tan^{-1}\frac{2\tau_{xy}}{\sigma_x - \sigma_y} = 90 - 22.5 = 67.5° \tag{7.32}$$

---

#### Arthur Casagrande (カーサグランデ, 1902〜1981)

オーストリア生まれのアメリカの土木工学者。1924年，今のウィーン工科大学土木工学科を卒業。父親の死後アメリカに渡り，MIT に職を求めに行ったとき会ったのがテルツァーギで，幸運なことに彼の私的な助手としてすぐ採用された。テルツァーギのもとで，US. Bureau や MIT の仕事をした。MIT にいるころ，液性限界試験装置を開発した。土質力学では，カーサグランデの分類法で知られる。

図6 カーサグランデ

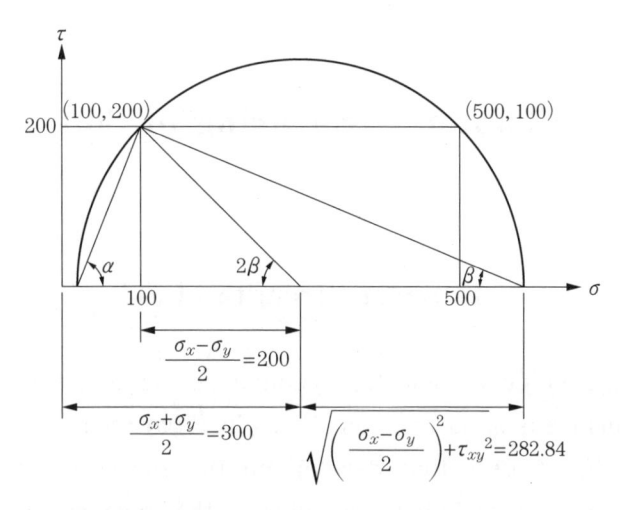

## 7.3 Mohr's Stress Circle Considering Shear Stresses

**Figure 7.8** Mohr's stress circle for Example 7.2
図7.8 モールの応力円

$$=\frac{500+100}{2}\pm\sqrt{\left(\frac{500-100}{2}\right)^2+200^2}=300\pm282.84 \text{ kN/m}^2 \quad (7.30)$$

より

$\sigma_1 = 582.84 \text{ kN/m}^2$

$\sigma_3 = 17.16 \text{ kN/m}^2$

となる。

また，主応力面の方向は

$$\beta=\frac{1}{2}\tan^{-1}\frac{2\tau_{xy}}{\sigma_x-\sigma_y}=\frac{1}{2}\tan^{-1}\frac{2\times200}{400}=22.5° \quad (7.31)$$

$$\alpha=\frac{\pi}{2}-\frac{1}{2}\tan^{-1}\frac{2\tau_{xy}}{\sigma_x-\sigma_y}=90-22.5=67.5° \quad (7.32)$$

となる。

# Chapter 8　Shearing of Soil

## 8.1　Shear Strength of Soil

When a soil sample is subjected to compressional loads, it fails along a shear surface (slip surface) as shown in **Figure 8.1**.

The stresses that develop on the shear surface are the normal stress $\sigma$ that is perpendicular to the shear surface and the shear stress $\tau$ that is parallel to the shear surface. The normal stress $\sigma$ causes particle crushing because stresses among soil grains result in failure of the individual grains. On the other hand, shear stress $\tau$ causes relative movement of the soil particles. In the case of soil failure, breakage of solid particles occurs because of the pressures created by the normal stress $\sigma$ and by the relative movement of soil particles caused by the shear stress $\tau$.

Shear failure on the shear surface arises even for small stresses because of the movement of soil particles. For this reason, in the case of soil failure, it is generally assumed that the failure occurs because of the shear stress and <u>not</u> because of the normal stress.

Accordingly, failure of a soil mass happens when the shear stress that develops in the ground from external loads exceeds the existing shear strength of the soil mass. Therefore, it is important to study the existing shear strength of soil when designing a structure in soil.

# 8章 土のせん断

## 8.1 土のせん断強度

土でできた供試体に上下方向から力を加えると，図8.1のようなせん断面（すべり面）が発生し壊れることがある。このせん断面に生じる応力は，せん断面に対して垂直な垂直応力 $\sigma$ と，平行なせん断応力 $\tau$ である。このうち，$\sigma$ は，せん断面の土粒子同士を押しつけて，粒子を壊そうとするのに働く応力で

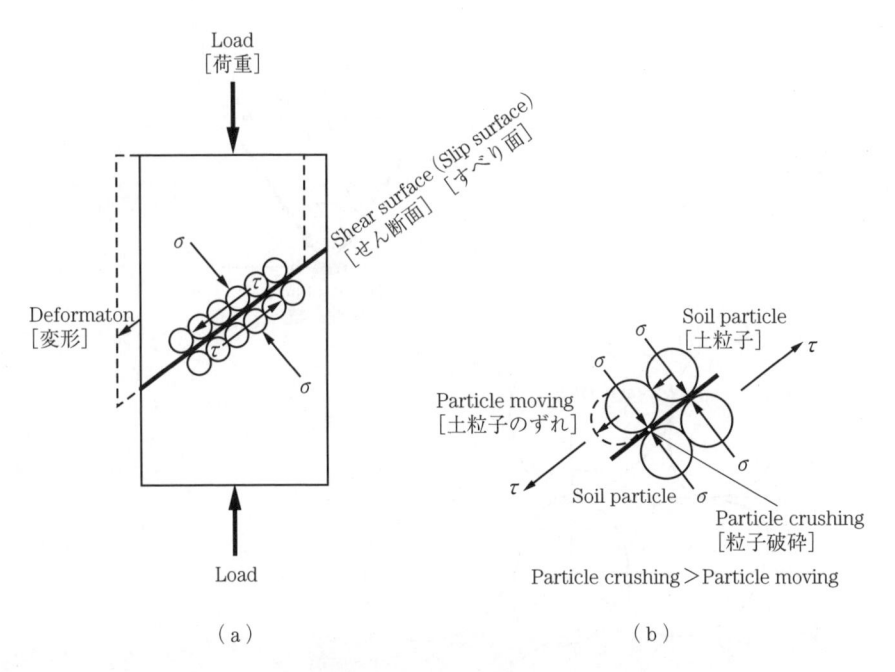

**Figure 8.1** Shear failure of soil
図8.1 土のせん断破壊

## 8.2 Coulomb's Failure Criterion

The classification of a soil depends on the size of the particles. A soil is classified as either coarse (gravel and sand) or fine (silt and clay). In studying the shear strength of soil, it is important to understand the behavior of coarse particles, such as sand, separately from the behavior of fine particles such as clay.

First, we present the behavior of a soil composed of large particles such as sand. The forces acting between soil particles are shown in **Figure 8.2**. Local failure can occur when a force $N$ acts at the contact point between soil particles. If the contact area is $a$, and the strength per unit area is $q$, then the force $N$ is given by the

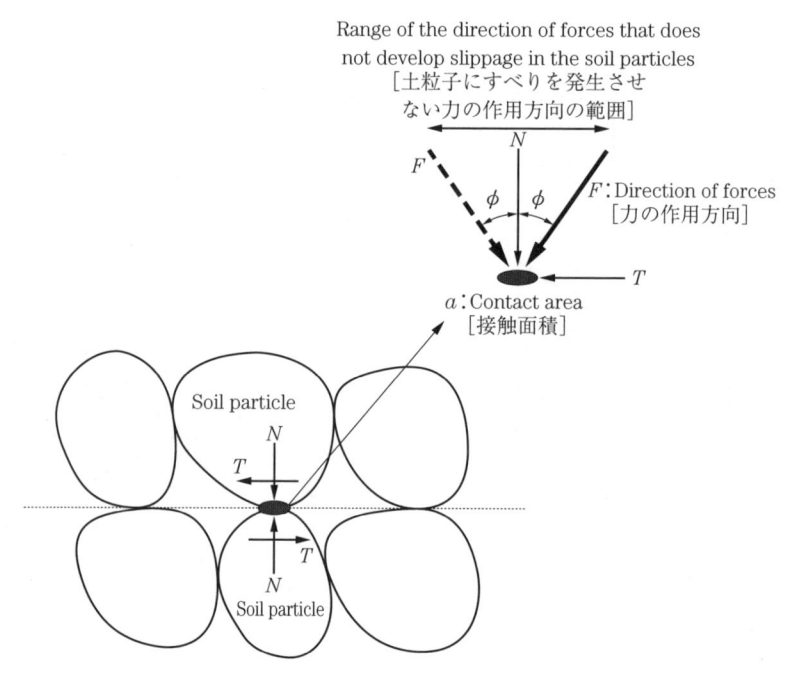

**Figure 8.2** Forces acting between soil particles
図 8.2 土粒子同士に作用する力

あるのに対し，せん断応力 $\tau$ は，せん断面の粒子を相対的に移動させようとする（ずらそうとする）のに働く応力である。土の破壊を考える場合，土粒子は硬いため $\sigma$ によって土粒子同士が押しつけられて壊れる粒子破壊より，$\tau$ によって土粒子同士が相対的に移動し，ずれを起こしてせん断面を発生させて壊れるせん断破壊の方が応力は小さくてすみ，容易であると考えられる。このため，土の破壊を考える場合，垂直応力ではなくせん断応力を対象として考えるのが一般的である。

したがって土の破壊は，外力の作用により地盤内に生じるせん断応力が，土の保持しているせん断強度より大きくなることにより発生すると考え，土構造物を設計する場合，土のもっているせん断強度を知ることが重要である。

## 8.2 クーロンの破壊基準

土は，粒径によって大きく粗粒分（礫分＋砂分）と，細粒分（シルト分＋粘土分）に分けられる。土のせん断強度を考える場合，粗粒分に相当する砂質土と，細粒分に相当する粘性土に分けて特性を考える必要がある。

まず，砂のような粒径の大きい土粒子の特性について考えてみる。図 8.2 は土粒子同士に作用する力の関係をみたものである。接触している土粒子に，$N$ の力が作用し局所的な流動が生じた場合，接触部の面積を $a$，接触部の単位面積あたりの強度を $q$ とすると

$$N = qa \tag{8.1}$$

となる。

また，接触部の単位面積あたりの摩擦抵抗が $s$ の場合，土粒子間の接触部の摩擦抵抗力 $T$ は

$$T = sa \tag{8.2}$$

となる。

ここで，土粒子接触部の摩擦抵抗と強度との比

following equation:

$$N = qa \tag{8.1}$$

If the frictional resistance per unit area is $s$, then the frictional force $T$ at the contact surface between the soil particles is given by:

$$T = sa \tag{8.2}$$

The ratio of the frictional resistance to the strength is:

$$\frac{T}{N} = \frac{s}{q} \tag{8.3}$$

Taking $T_{max}$ as the limiting value of the frictional resistance of the contact surfaces between soil particles, then if

$$\frac{T_{max}}{N} \geq \frac{T}{N} : \text{No slippage of the soil particles occurs}$$

$$\frac{T_{max}}{N} < \frac{T}{N} : \text{Slippage of the soil particles occurs}$$

Here, $T/N$ can be calculated from the condition of the contact between the soil particles and the material properties of the contact area between the soil particles. $T/N$ is a parameter and is taken to be $(\tan\phi)$. $\phi$ is the angle of internal friction.

When $N$ and $T$ act on a ground of cross-sectional area $(A)$, the normal stress $\sigma$ and shear stress $\tau$ can be calculated as follows:

$$\sigma = \frac{N}{A} \tag{8.4}$$

$$\tau = \frac{T}{A} \tag{8.5}$$

From Equations (8.3), (8.4) and (8.5); the following equations are obtained:

$$\frac{\tau}{\sigma} = \frac{T}{N} = \tan\phi \tag{8.6}$$

$$\tau = \sigma \tan\phi \tag{8.7}$$

Equation (8.7) gives the shear strength $\tau$ of a soil containing large particles such as sand.

## 8.2 Coulomb's Failure Criterion

$$\frac{T}{N} = \frac{s}{q} \tag{8.3}$$

を考えてみる。

粒子接触部の摩擦抵抗の限界値が $T_{max}$ であるとすると

$\dfrac{T_{max}}{N} \geq \dfrac{T}{N}$：土粒子同士はすべらない

$\dfrac{T_{max}}{N} < \dfrac{T}{N}$：土粒子同士がすべる

と考えられる。

ここで，$T/N$ は土粒子間の接触部の物性や粒子の接触状態によって求まる定数と考えることができるため，$T/N$ を $\tan\phi$ とおき，$\phi$ を内部摩擦角と呼ぶ。

$N$，$T$ の力が面積 A の地盤内に作用する場合，垂直応力 $\sigma$，せん断応力 $\tau$ は

$$\sigma = \frac{N}{A} \tag{8.4}$$

$$\tau = \frac{T}{A} \tag{8.5}$$

であり，
式(8.3)〜(8.5)から

$$\frac{\tau}{\sigma} = \frac{T}{N} = \tan\phi \tag{8.6}$$

$$\tau = \sigma \tan\phi \tag{8.7}$$

が得られる。

式(8.7)は，砂のような粒径の大きい土のせん断強度 $\tau$ を表す式である。

つぎに，粘土のような粒径の小さい土粒子の特性について考えてみる。粘土粒子は，薄板構造で粒径が5μm以下と小さく，表面は負に帯電し粘土粒子自体は負電荷の粒子として振る舞う特徴をもっている。このため，土中に水分があると，水分子を構成する $H^+$ と粘土粒子表面の $O^{-2}$ が強く結合し，水分子が粘土粒子に吸着されて水膜（吸着水層）を形成する（**図8.3**）。このため，粘

Next, we examine the behavior of soil composed of small particles such as clay. Clay particles are thin-walled or platy-structures with dimensions often less than 5 μm. In addition, the surfaces of clay are negatively charged causing the clay particle to act like a negatively charged ion. When water is present in soil, the positive ion $H^+$ of water can combine strongly with the negative ion $O^{-2}$ of the clay particle. The positive ions are adsorbed by the clay particles forming a film of water around and between the particles (**Figure 8.3**) (i.e., adsorption water layer). Because of this phenomenon, the clay particles are bound together as well as separated by the adsorption water (Figure 8.3). As a result of this binding force by water, clay particles possess strength (viscosity) that tends to resist displacement. This characteristic of clay particles is called cohesion $c$.

In nature, soil consists of various particles sizes, such as, clay, sand and gravel, etc. Thus, when determining the shear strength of soil, it is important to consider the effects of stress (e.g., the resistance against displacement at the contact between large particles such as sand) as well as the cohesion from the binding force between small particles like clay.

The equation for shear strength of a natural soil is given by:

$$\tau = c + \sigma \tan\phi \tag{8.8}$$

Equation (8.8) is known as Coulomb's failure criterion,

where, $\tau$ : Shear strength, $c$ : Cohesion, $\sigma$ : Normal stress, $\phi$ : Angle of internal friction.

Adsorption water does not form in sand because the sizes of the particles are too large. Thus, the cohesion is taken as zero, i.e. ($c=0$) for sand. Therefore, the shear strength is given by the following equation.

$$\tau = \sigma \tan\phi \tag{8.9}$$

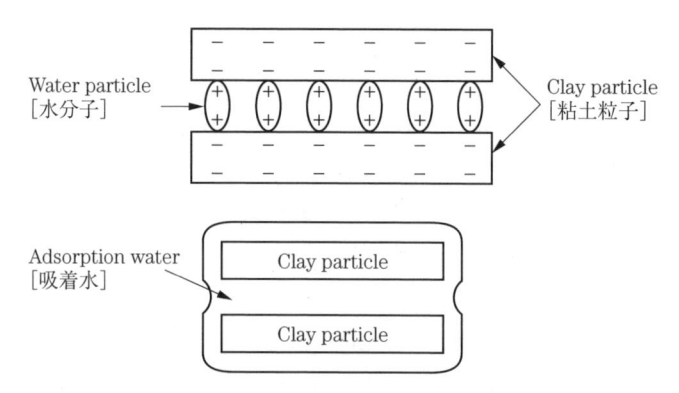

**Figure 8.3** Clay particles and adsorption water layer
図 8.3　粘土粒子と吸着層

　土粒子同士は，水膜を介して接触し，このたがいの結合力に由来するずれに対して抵抗しようとする強度（粘性）をもっており，これを粘着力 $c$ と呼んでいる。

　ところで，一般的な土は，粘土から砂，礫まで種々の粒径が存在するため，土のせん断強度を考える場合，粒子の大きい砂のような土粒子の接触部のずれに対する抵抗力と，粒子の小さい粘土のような土粒子間の結合力に起因する粘着力を併せたものとして考える必要がある。

　このため，一般的な土のせん断強度を表す式は

$$\tau = c + \sigma \tan\phi \tag{8.8}$$

で表され，これをクーロンの破壊基準という。

　ここで，$\tau$：せん断強度，$c$：粘着力，$\sigma$：垂直応力，$\phi$：内部摩擦角である。

　土のせん断強度は，砂では粒子が大きく粒子同士が吸着層を形成できないため $c = 0 \text{ kN/m}^2$ であり

$$\tau = \sigma \tan\phi \tag{8.9}$$

で表される。

　また，粘土では土粒子間の接触部のずれに対する抵抗力の影響がないため $\phi = 0°$ であり

On the other hand, clay soil has no resistance ($\phi=0°$) against displacement at the contact between particles. Therefore, for clays, the following equation for shear strength is obtained:

$$\tau = c \tag{8.10}$$

When working with soil, it is also necessary to consider the influence of pore water pressure along with the total stress and the effective stress.

The total stress is given by Equation (8.8), Coulomb's failure criterion when using effective stress is derived as follows:

from $\sigma' = \sigma - u$, we obtain,

$$\tau = c' + (\sigma - u)\tan\phi = c' + \sigma'\tan\phi' \tag{8.11}$$

where $u$ is the pore water pressure

For soil with a large coefficient of permeability, like sand that permits drainage during shear, excess pore water pressure does not develop. Thus, Equation (8.11) reduces to

$$\tau = \sigma\tan\phi = \sigma'\tan\phi_d \tag{8.12}$$

**[Example 8.1]** Direct shear tests were performed on a soil specimen with a diameter of 5 cm for a range of normal stresses. The results obtained from the tests are given in **Table 8.1**. Calculate the cohesion $c$ and the angle of internal friction $\phi$ for this soil.

**Solution**  The normal and shear stresses are calculated by dividing the normal loads and shear forces by the cross-sectional area of shear surface and are given in **Table 8.2**.

Based on the results given in Table 8.2, the resulting relationship between the normal stresses and shear stresses is shown in **Figure 8.4**. According to Coulomb's failure criterion, a straight line is drawn through all of the points. The $y$-axis intercept is the cohesion $c$ and the slope of the line is the angle of internal friction $\phi$. Thus, $c$ and $\phi$ are:

$c = 10 \text{ kN/m}^2$
$\phi = 26.6°$

$$\tau = c \tag{8.10}$$

で表される。

また，土を扱う場合には，間隙水圧の影響を考慮し，全応力と有効応力に分けて扱う必要がある。式(8.8)は，全応力における表現であるが，有効応力に対するクーロンの破壊基準は $\sigma' = \sigma - u$ より

$$\tau = c' + (\sigma - u)\tan\phi = c' + \sigma' \tan\phi' \tag{8.11}$$

で表される。

ここで，$u$：間隙水圧である。

透水係数が大きくせん断中に排水が可能な砂のような材料の場合，過剰間隙水圧の発生がないため $u=0$ となり

$$\tau = \sigma \tan\phi = \sigma' \tan\phi_d \tag{8.12}$$

で表される。

【例題 8.1】 直径 5 cm の土の供試体に垂直荷重を作用させた後，水平変位を加えてせん断を行い，表 8.1 の実験結果を得た。この土の粘着力 $c$ および内部摩擦角 $\phi$ を求めよ。

Table 8.1  Test results
表 8.1  実験結果

| Normal loads [垂直荷重] $N$ | 100 | 200 | 400 |
|---|---|---|---|
| Shear forces [せん断力] $S$ | 69.63 | 119.63 | 217.68 |

**解答**　垂直荷重およびせん断力をそれぞれ断面積で割り，垂直応力，せん断応力を求めると，表 8.2 のようになる。

表 8.2 で求まった結果を基に，垂直応力-せん断応力の関係を求めると，図 8.4 になる。クーロンの破壊基準を考慮し，各点を通る直線から $c$, $\phi$ を求めると

$c = 10$ kN/m$^2$

$\phi = 26.6°$

となる。

Chapter 8  Shearing of Soil

**Table 8.2** Calculated results
**表 8.2** 計算結果

| Normal loads<br>［垂直荷重］$N$ | Shear force<br>［せん断力］$S$ | Normal stress<br>［垂直応力］［$kN/m^2$］ | Shear stress<br>［せん断応力］［$kN/m^2$］ |
|---|---|---|---|
| 100 | 69.63 | 50.96 | 35.48 |
| 200 | 119.63 | 101.91 | 60.96 |
| 400 | 217.68 | 203.82 | 111.92 |

Cross-sectional area ［断面積］：19.625 $cm^2$

---

### Osborne Reynolds （レイノルズ，1842〜1912）

イギリスの物理学者。1867年ケンブリッジ大学卒業。翌年，マンチェスターのオーウェンス大学の工学の教授となった。イギリスにおいては工学の教授「Professor of Engineering」の称号は歴史上初めてで，しかも，財源はマンチェスターの製造工業から出されたものであった。いまでいう企業による寄付口座というところか。

土質力学ではレイノルズ数，限界流速で知られる。

図7　レイノルズ

8.2 Coulomb's Failure Criterion　45

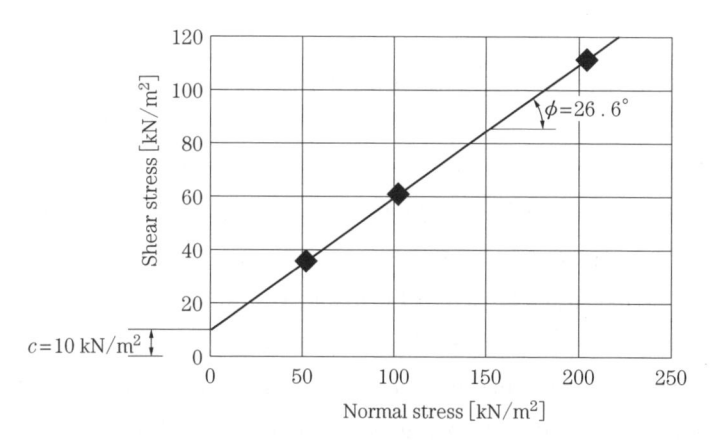

**Figure 8.4** Relationship between shear stress and normal stress from the example
図 8.4　せん断応力と垂直応力の関係

---

### Joseph Valentin Boussinesq（ブシネスク，1842〜1929）

　フランスの数学者，物理学者。「ブシネスクの近似」で有名。熱力学，振動，光と熱に貢献した。
　「乱流」という言葉は，ブシネスクに負うところが大である。津波はよく知られているが，これは海での孤立波であって，J. S. Russell が発見した運河で発生した孤立波現象を，数学的に解析したのがブシネスクである。
　土質力学では，半無限媒質に集中荷重が加わった場合の地盤内応力で有名。また，サンブナンの原理で有名なサンブナンの弟子である。

図8　ブシネスク

## 8.3 Mohr-Coulomb's Failure Criterion

Consider a ground subjected to maximum $\sigma_1$ and minimum $\sigma_3$ principal stresses. A Mohr's circles for a gradual increase in maximum principal stress $\sigma_1$ ($\sigma_{1-1} \rightarrow \sigma_{1-2} \rightarrow \sigma_{1-3}$) but with a constant minimum principal stress $\sigma_3$ is shown in **Figure 8.5**. The Mohr's circle increases in diameter as $\sigma_1$ increases. The shear strength of soil is given by the Coulomb's failure criterion. The shear failure of soil occurs when the stress exceeds point (A) on the line of Coulomb's failure criterion. A soil cannot support stresses above the line representing the Coulomb's failure criterion (Figure 8.5). The stress circle that corresponds to the failure stress $\sigma_{1-3}$ is known as the failure stress circle. The straight line of Coulomb's failure criterion connecting the failure stress circles is called the failure envelope.

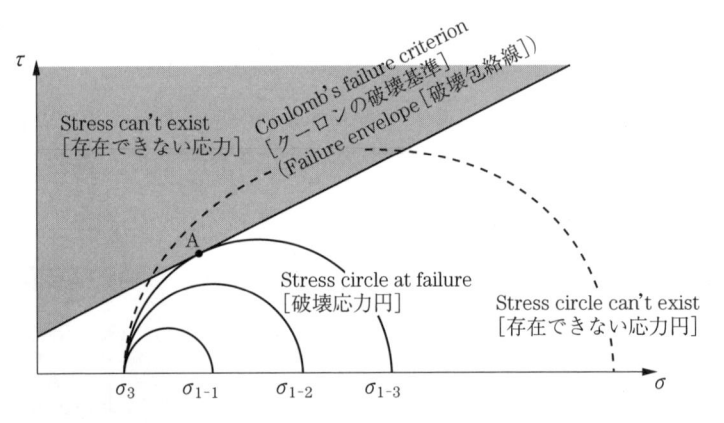

**Figure 8.5** Mohr-Coulomb's failure criterion
図 8.5 モール・クーロンの破壊基準

It is necessary to evaluate different stress conditions because the stress in the ground varies with depth. **Figure 8.6** shows failure stress circles that correspond to different depths in the ground ($h_1$, $h_2$,

## 8.3 モール・クーロンの破壊基準

地盤内に主応力 $\sigma_1$, $\sigma_3$ が作用するとき，$\sigma_3$ を一定として $\sigma_1$ を徐々に増加（$\sigma_{1-1} \rightarrow \sigma_{1-2} \rightarrow \sigma_{1-3}$）させたときのモールの応力円を描くと，モール円は図 8.5 に示すように $\sigma_1$ が大きくなるに従い大きくなる。しかし，土のせん断強度はクーロンの破壊基準で示され，クーロンの破壊基準に接する A 点を超えると土はせん断破壊を起こすため，これ以上の応力は存在できないこととなる。このとき，この破壊時の $\sigma_{1-3}$ に相当する応力円を破壊応力円といい，この破壊応力円に接するクーロンの破壊基準の直線を破壊包絡線という。

ところで，地盤内では深度などによって応力が変化するため，種々の応力状態に対する検討が必要となる。図 8.6 のように地盤内で深さが異なる地点

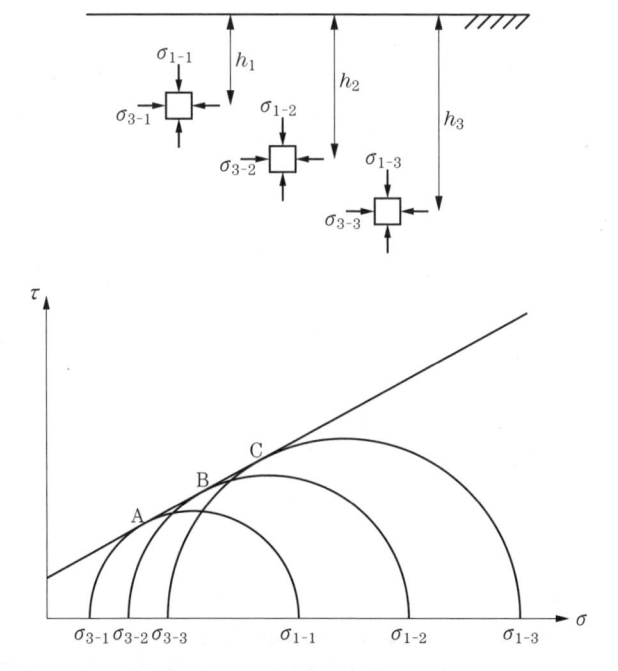

**Figure 8.6** Stress circles at failure corresponding to various principal stresses
図 8.6　いくつかの主応力に対する破壊応力円

$h_3$). Points (A), (B) and (C) on each stress circle fulfill the failure conditions. A line joining all of these points is the "failure envelope" satisfying the "general failure criterion" for all the circles. This failure criterion is called the "Mohr-Coulomb's failure criterion".

Now, we develop the Mohr-Coulomb's failure criterion by considering the principal stresses. From **Figure 8.7**, the radius of the Mohr's stress circle is given by:

$$\overline{AE} = \overline{CE} = \frac{\sigma_1 - \sigma_3}{2} \tag{8.13}$$

From $\overline{CE} = \overline{CD} + \overline{DE}$ (8.14)

and

$$\overline{CD} = c \cos\phi \tag{8.15}$$

$$\overline{DE} = \frac{\sigma_1 + \sigma_3}{2} \sin\phi \tag{8.16}$$

which results in

$$\frac{\sigma_1 - \sigma_3}{2} = c \cos\phi + \frac{\sigma_1 + \sigma_3}{2} \sin\phi \tag{8.17}$$

For the case of sand, ($c = 0$ kN/m²). Hence, from Equation (8.17), the following equations are obtained:

$$\sin\phi = \frac{\sigma_1 - \sigma_3}{\sigma_1 + \sigma_3} \tag{8.18}$$

or

$$\frac{\sigma_1}{\sigma_3} = \frac{1 + \sin\phi}{1 - \sin\phi} \tag{8.19}$$

Expression of stress conditions using Mohr's stress circle is an effective method but becomes complicated when several stress circles need to be drawn for various stress conditions. To overcome this complication, an elegant method ($p-q$ plot) equivalent to Mohr's stress circle is used. The ($p$, $q$) on Mohr's stress circle is the coordinate of point (A) as shown in **Figure 8.8**, where $p$ is the center of the stress circle and $q$ is the radius of the Mohr's circle (maximum principal

## 8.3 Mohr-Coulomb's Failure Criterion

($h_1$, $h_2$, $h_3$) での主応力に対する破壊応力円を描くと，各応力円上で破壊条件を満足する点 A, B, C が存在し，その点を連ねた接線は，すべての応力円に共通な破壊基準を満足する破壊包絡線であると考えられる。このような破壊基準をモール・クーロンの破壊基準という。

ここで，モール・クーロンの破壊基準を主応力によって表してみる。

図 8.7 より，モールの応力円の半径は

$$\overline{AE} = \overline{CE} = \frac{\sigma_1 - \sigma_3}{2} \tag{8.13}$$

である。

$$\overline{CE} = \overline{CD} + \overline{DE} \tag{8.14}$$

$$\overline{CD} = c \cos\phi \tag{8.15}$$

$$\overline{DE} = \frac{\sigma_1 + \sigma_3}{2} \sin\phi \tag{8.16}$$

より

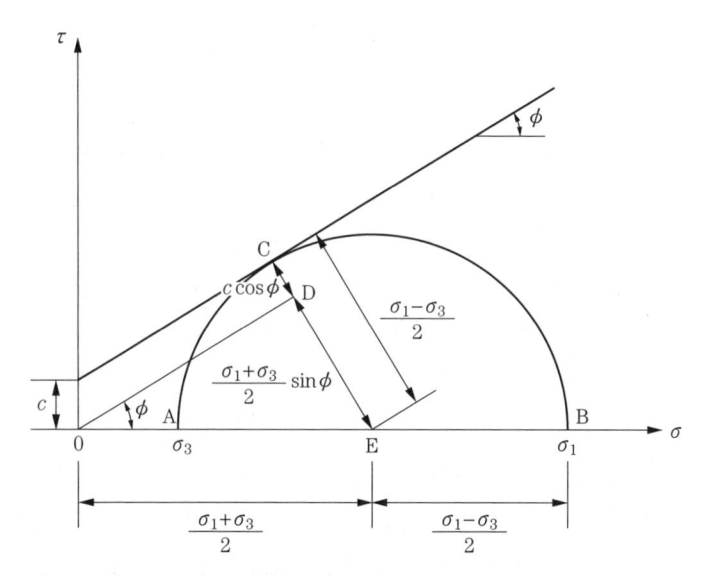

**Figure** 8.7  Expression of Mohr-Coulomb's failure criterion considering principal stresses
図 8.7 モール・クーロンの破壊基準の主応力での表現

50　Chapter 8　Shearing of Soil

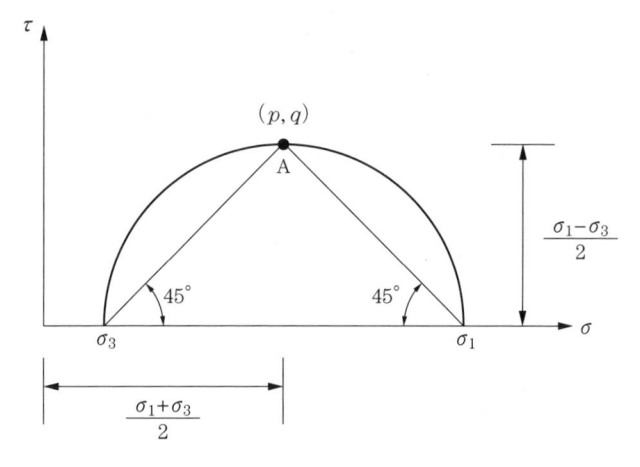

**Figure 8.8** Mohr's stress circle and ($p$, $q$)
図 8.8　モールの応力円と $p$, $q$

stress).

$$p = \frac{\sigma_1 + \sigma_3}{2} \tag{8.20}$$

$$q = \frac{\sigma_1 - \sigma_3}{2} \tag{8.21}$$

Lines at 45° on both the left and right sides from point (A) are drawn as shown in Figure 8.8. The intersections of these lines with the horizontal axis are coordinates of $\sigma_1$ and $\sigma_3$. The ($p$-$q$) coordinates are used to show the change in stress conditions. An example of this is shown in **Figure 8.9** (such as A → B → C). A change in the stress conditions of soil is called the soil's "stress history". A line showing the locus of points on the ($p$-$q$) plot is called the "stress path".

【Example 8.2】 A cylindrical soil specimen, 5 cm in diameter, was loaded under an axial stress of 300 kN/m² and a horizontal stress (radial stress) of 100 kN/m². The specimen failed at an angle of 50° with the horizontal. Calculate the cohesion $c$ and the angle of internal friction $\phi$. Assume that the shear stress did not develop around the

## 8.3 Mohr-Coulomb's Failure Criterion

$$\frac{\sigma_1-\sigma_3}{2}=c\cos\phi+\frac{\sigma_1+\sigma_3}{2}\sin\phi \tag{8.17}$$

となる。

砂の場合，$c=0\,\mathrm{kN/m^2}$ より式(8.17)は

$$\sin\phi=\frac{\sigma_1-\sigma_3}{\sigma_1+\sigma_3} \tag{8.18}$$

あるいは

$$\frac{\sigma_1}{\sigma_3}=\frac{1+\sin\phi}{1-\sin\phi} \tag{8.19}$$

で表される。

　モールの応力円は応力状態を表現するのに有効な方法であるが，種々の応力を扱う場合いくつもの応力円を描く必要があり，図が込み入ってくる。このため，モールの応力円と等価であるすっきりした表現法として $p$-$q$ プロットがある。図 8.8 に示すように，モールの応力円上の $(p, q)$ は，A 点の座標値となり，このとき，$p$ は応力円の中心，$q$ はモール円の半径（最大せん断応力）である。

$$p=\frac{\sigma_1+\sigma_3}{2} \tag{8.20}$$

$$q=\frac{\sigma_1-\sigma_3}{2} \tag{8.21}$$

また，A 点から左右に 45°の線を引き，横軸との交点の座標は $\sigma_1$, $\sigma_3$ である。応力状態の変化を $p$-$q$ 座標で表すと，応力は図 8.9 のように A → B → C

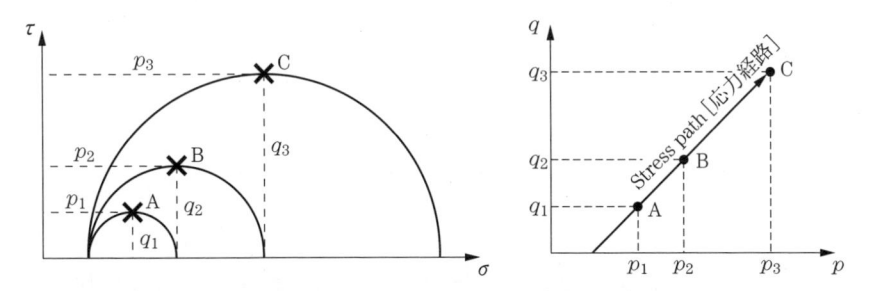

**Figure 8.9** Mohr's stress circle and stress path
図 8.9　モールの応力円と応力経路

specimen.

**Solution**　From the minimum and maximum stresses, the principal stresses are:
$$\sigma_1 = 300 \text{ kN/m}^2$$
$$\sigma_3 = 100 \text{ kN/m}^2$$
The Mohr's stress circle using the principal stresses is shown in **Figure 8.10**. A line for Coulomb's failure criterion connecting Mohr's circle with an angle 50° for the failure surface is drawn. The $c$ and $\phi$ are obtained as follows:
$$c = 65.27 \text{ kN/m}^2$$
$$\phi = 10°$$

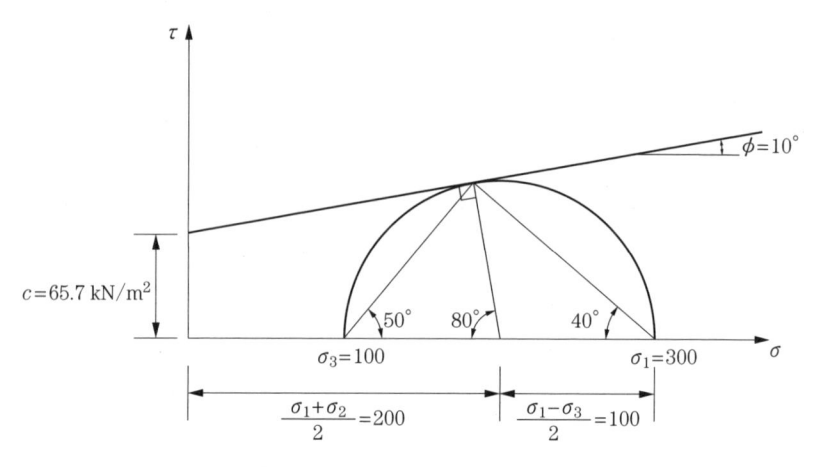

**Figure 8.10**　Mohr's stress circle
図 8.10　モールの応力円

と移動し，この土中内の応力状態の変化を応力履歴といい，$p$–$q$ 上の軌跡として表した線を応力経路と呼ぶ。

【例題 8.2】 直径 5 cm の土の円筒供試体の垂直方向に 300 kN/m² の応力を，水平方向に 100 kN/m² の応力（拘束圧）を作用させたとき，供試体が水平面と 50° の角度で破壊した。このときの粘着力 $c$ と内部摩擦角 $\phi$ を求めよ。なお，供試体周面にはせん断応力の発生はないとする。

**解答** 主応力は，応力の大小より
$\sigma_1 = 300 \text{ kN/m}^2$
$\sigma_3 = 100 \text{ kN/m}^2$
である。
 主応力を基にモールの応力円を描くと**図 8.10** となる。破壊面の角度が 50° であることを考慮して，モール円に接するクーロンの破壊基準を描き，$c$, $\phi$ を求めると
$c = 65.27 \text{ kN/m}^2$
$\phi = 10°$
となる。

---

### Albert Atterberg（アタバーグ，1846〜1916）

スウェーデンの化学者であるが，地質工学者として知られる。1872 年ウプサラ大学で Ph.D. を取り，そこで講師まで勤めた。1891〜1900，カルマの大学でカラスムギとトウモロコシの分類で，農学に関し多くの論文を書いた。
 アタバーグのコンシステンシー限界で土質力学では知られている。

図 9　アタバーグ　　図 10　液性限界測定装置

## 8.4 Shear Properties of Sand and Clay

A soil is composed of particles of various sizes and is classified based on the size of the particles. Depending on the particle sizes, a soil is classified as coarse (gravel and sand) or fine (silt and clay). In considering the shear strength of a soil, the behavior of a coarse soil (e. g., sandy soil) is often studied separately from the behavior of a fine soil (clayey soil). For sandy soil, it is important to study the dilatancy properties (change in volume) during shear deformation.

Let's examine the shear deformation of two soils, one that is loose condition ( a ) and other that is dense condition ( b ) as shown in **Figure 8.11**. When shear stress is applied to a loose soil with a large initial void ratio, the particles are rearranged due to the movement of particles during shear. The particles tend to slide into the large void area between particles causing a decrease in bulk volume and an increase in density. On the other hand, when shear stress is applied to a dense soil with a small initial void ratio, the particles tend to slide up and onto the top of the neighboring particles. After shearing, an initially densely soil will increase in volume and decrease in density. Whether a sandy soil will increase or decrease in volume depends on the initial void ratio of the soil. This phenomenon that causes a change in the volume during shearing is called "dilatancy". For dilatancy, a decrease in the volume is taken as negative (−) and an increase in the volume is taken as positive (+).

Shear stress-displacement relationships and normal stress-displacement relationships for dense and loose sand are shown in **Figure 8.12**. For a loose sand subjected to an increasing shear stress, the shear stress is almost constant after a certain amount of shear displacement. The change in normal displacement with increasing

## 8.4 砂と粘土のせん断特性

　土は種々の粒子が集まってできており，粒径を基に大きく粗粒分（礫分＋砂分），細粒分（シルト分＋粘土分）に分けられ，その特性は粒径によって区分される。土のせん断特性を考える場合，粗粒分に相当する砂質土と細粒分に相当する粘性土に分けて考える必要がある。

　砂質土のせん断特性を考える場合，せん断変形に伴い体積が変化するダイレイタンシー特性が重要となる。ここで，図 8.11 に示すような（a）緩詰め，（b）密詰めの砂地盤におけるせん断変形を考えてみる。

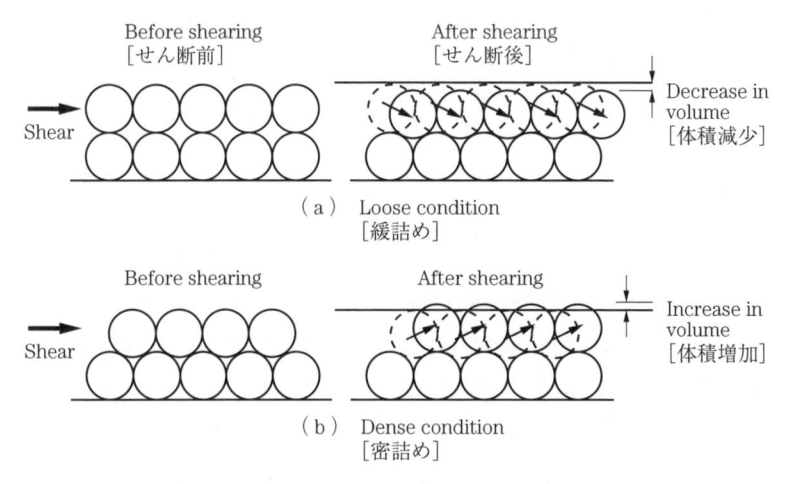

**Figure 8.11** The change in the volume during shear deformation
図 8.11　せん断による体積変化の様子

　初期間隙比が大きく密度が緩い状態においてせん断を行うと，粒子間の隙間が大きいため，せん断による粒子の移動によって隙間に粒子が落ち込み，密度は緩い状態から密な状態に変化し体積は減少する。一方，初期間隙比が小さく密度が密な状態においてせん断を行うと，粒子間に隙間がないため，せん断により粒子は隣の粒子の上にすべり上がり，密度は密な状態から緩い状態に変化して体積が増加する。

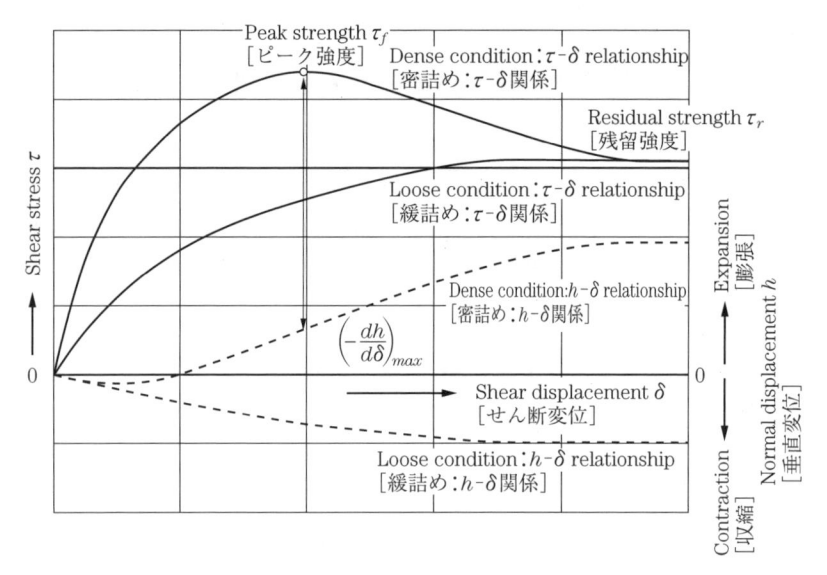

**Figure 8.12** Shear stress-displacement relationships and volume change for a sandy soil
図 8.12 砂質土におけるせん断変位とせん断応力および体積変化との関係

shear stress is negative, indicating a contraction (i.e., decrease in volume). In contrast, for a dense sand subjected to peak shear strength, the shear stress decreases after the peak value. After strain-softening, a constant residual strength remains. In this case, the change in normal displacement is positive indicating an expansion (i.e., increase in volume). Also, for a dense sand, the displacement that occurs at peak strength coincides with a displacement of $\left(-\dfrac{dh}{d\delta}\right)_{max}$, that is the maximum inclination (gradient of slope) of normal displacement.

For clayey soil, the consolidation history of the soil needs to be considered (e.g., normally consolidated or over-consolidated, etc.). Relationships among vertical strain $\varepsilon_1$, $q$ and volumetric strain $\varepsilon_v$ for normally consolidated clays and over-consolidated clays are shown in **Figure 8.13**. For the normally consolidated condition (where

8.4 Shear Properties of Sand and Clay    57

このように，砂質土では地盤の間隙比によってせん断による体積変化に違いがみられる。このようなせん断によって体積が変化する現象をダイレイタンシーと呼ぶ。ダイレイタンシーは，体積が減少する場合を負（－），体積が増加する場合を正（＋）とする。

図 8.12 に，密詰め，緩詰めの砂地盤におけるせん断変位とせん断応力および垂直変位との関係を示す。緩詰め地盤では，せん断応力が増加した後，あるせん断変位以降ほぼ一定値のせん断応力を示すとともに，垂直変位の変化は負のダイレイタンシーである収縮を示す。これに対し，密詰め地盤では，せん断

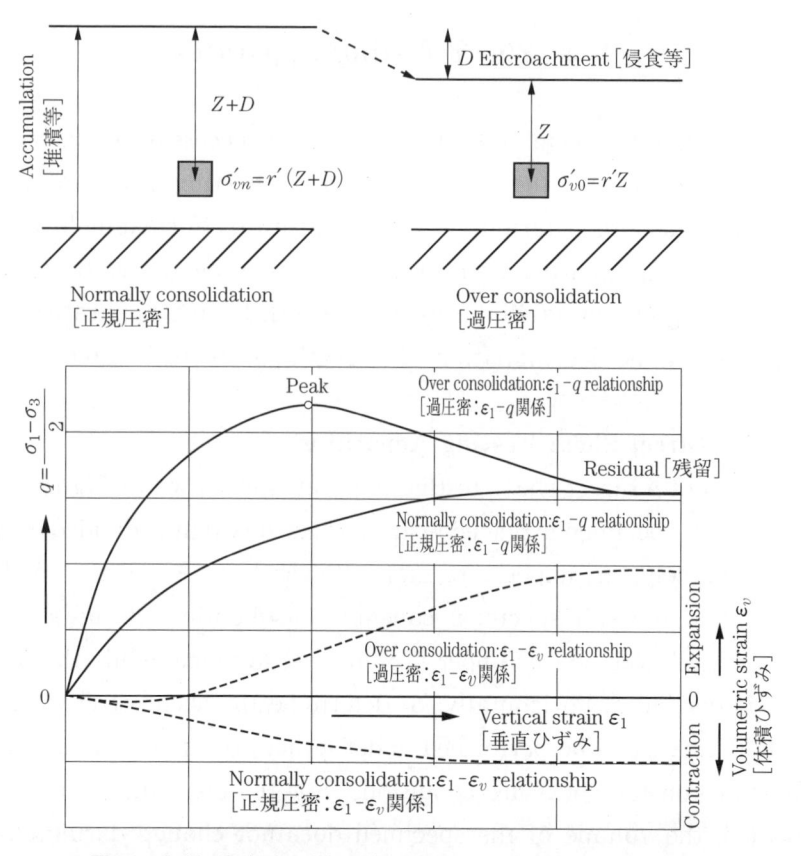

**Figure 8.13** Relationships of $\varepsilon_1$ with $q$ and $\varepsilon_v$ for clayey soil
図 8.13 粘性土における $\varepsilon_1$ と $q$，$\varepsilon_v$ との関係

the existing effective stress is equal to the maximum effective stress), the shear stress is almost constant after a certain amount of shear displacement, similar to that was observed for a loose sand. In this case, the volumetric strain $\varepsilon_v$ is contraction. In contrast, for the over-consolidated condition (where the existing effective stress is smaller than the previous maximum effective stress) after the peak strength has been reached and strain-softening has occurred, a constant residual strength remains. In this case, the volumetric strain $\varepsilon_v$ is expansion.

## 8.5 Shear Testing Apparatus

To determine the shear strength of a soil, it is necessary to determine the cohesion $c$ and the angle of internal friction $\phi$ according to Coulomb's failure criterion. Several apparatuses (for example, direct shear testing apparatus, unconfined compressive strength testing apparatus and triaxial compressive strength testing apparatus, etc.) are used for the determination of strength parameters in the laboratory.

(1) **Direct Shear Testing Apparatus**

A sketch of a direct shear testing apparatus is shown in **Figure 8.14** (a). A soil specimen is placed inside a shear box that is divided in two parts, namely, an upper box and a lower box. A normal load is applied to the soil specimen causing consolidation of the sample. Either the lower box or upper box is fixed in place, while the other box is displaced, horizontally, to determine the shear strength of a soil. The test can be performed under either constant volume conditions or constant pressure conditions. For the constant volume test method, the volume of the specimen does not change during shear because the normal stress is controlled to maintain a constant normal

応力が増加してピーク強度を示した後低下し，ひずみ軟化を示した後，最終的に一定の残留強度を示すとともに，垂直変位の変化は正のダイレイタンシーである膨張を示す。また，密詰め地盤においてピーク強度の変位と垂直変位の勾配が最大となる $\left(-\dfrac{dh}{d\delta}\right)_{max}$ の変位は一致すると考えられる。

粘性土のせん断特性を考える場合，正規圧密，過圧密などの圧密履歴が問題となる。図8.13に正規圧密，過圧密での垂直ひずみ $\varepsilon_1$，$q$，体積ひずみ $\varepsilon_v$ の関係を示す。

現在までに受けた最大の有効応力が，現在の有効応力と等しい正規圧密状態では，緩詰め砂と同様，せん断応力が増加した後，あるせん断ひずみ以降ほぼ一定値を示し，体積ひずみ $\varepsilon_v$ は収縮を示す。これに対し，現在の有効応力が過去に受けた最大の有効応力よりも小さい過圧密状態では，密詰め砂と同様，ピーク強度を示した後低下するひずみ軟化の後，最終的に残留強度を示すとともに，体積ひずみ $\varepsilon_v$ は膨張を示す。

## 8.5　せん断試験機

土のせん断強度を求める場合，クーロンの破壊基準（$\tau = c + \sigma \tan\phi$）にお

---

**Henry Darcy**（ダルシー，1803〜1858）

フランスの工学者。フランス・東部のディジョンに生まれ，1821年，パリのエコール・ポリテクニークに入学，2年後 École de Ponts et Chaussées に移った。後，軍隊の工兵に入った。工兵としてディジョン周辺の水道網を作り，1848年には Côte-d'Or 地方のチーフエンジニアとなり，ついでパリにある水道・舗装道路オフィスの部長に昇進した。この間，ピトー管やダルシーの法則となる実験を重ねた。1858年パリへ行く途中肺炎で死亡，ディジョンに埋葬された。透水性の単位・ダルシーはよく知られている。

図11　ダルシー

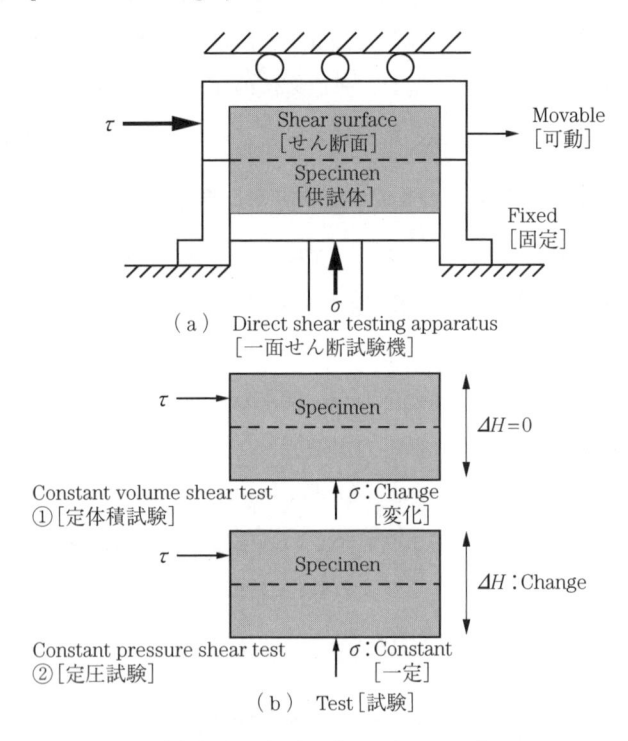

**Figure 8.14** (a) A sketch of a direct shear testing apparatus and (b-①) constant volume and (b-②) constant pressure conditions
図 8.14　一面せん断試験機の概要

deformation (Figure 8.14 (b-①)). For the constant pressure test method, the normal stress is kept constant during shearing (Figure 8.14 (b-②)).

　　　Shear tests are performed for several normal loads. After consolidation, a shear test is carried out for each desired normal load, and the shear stress and shear displacement are calculated under each normal load. For a constant volume test, a stress path is drawn based on the results obtained for each normal stress as shown in **Figure 8.15** ( a ). From the line of the stress path, the intercept in the $\tau$ axis is taken as the cohesion $c$ and the slope of the line is taken as the angle

ける粘着力 $c$ および内部摩擦角 $\phi$ を求めることが必要となる。室内においてこれら土の強度定数を求めるための試験機として，一面せん断試験，一軸圧縮試験，三軸圧縮試験などがある。

### （1） 一面せん断試験

図 8.14 に一面せん断試験機の概要を示す。これは上下 2 段で構成された箱の中に供試体を入れ，供試体に垂直荷重を加えて圧密を行った（図 8.14(a)）後，上下箱のどちらか一方を固定して他方を水平に移動させることで，上下箱

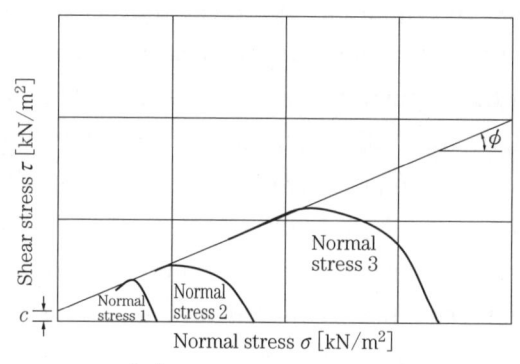

（a） A constant volume shear test

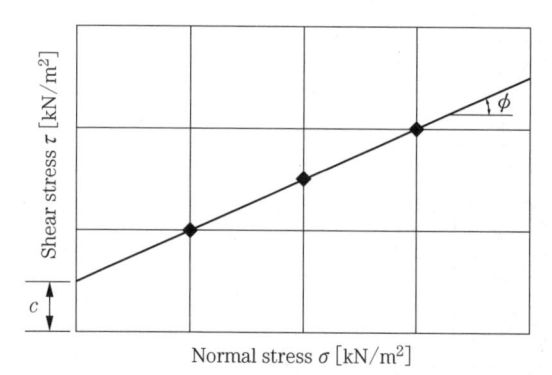

（b） A constant pressure shear test

**Figure 8.15** Relationship between normal and shear stresses for (a) constant volume and (b) constant pressure tests
図 8.15 垂直応力とせん断応力の関係

of internal friction $\phi$. For a constant pressure test method, shear strengths corresponding to each normal stress are plotted as shown in Figure 8.15 ( b ). Remember, in the constant pressure test method, the normal stress is held constant at each desired load. $c$ and $\phi$ are determined from the straight line that connects these points.

( 2 ) **Unconfined Compressive Strength Test**

A sketch of an unconfined compressive strength test is shown in **Figure 8.16**. In this test method, the compressive strength is determined by using a cylindrical specimen that is subjected to axial compression and that is unrestricted (i.e., unconfined or zero radial stress) around the circumference of the sample. The maximum compressive strength obtained during an unconfined compressive test is called the unconfined compressive strength $q_u$. This test can be easily performed because no radial stress is required. However, this test cannot be readily performed on sand because a sand specimen cannot maintain its integrity (i.e., maintain its shape or stand) without a radial stress.

Compressive stress
［圧縮応力］

Specimen

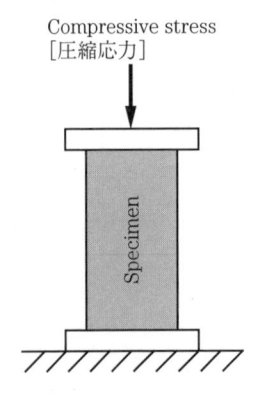

**Figure** 8.16 An outline of an unconfined compressive strength test
図 8.16 一軸圧縮試験機の概要

Results obtained from unconfined compressive strength tests are shown in **Figure 8.17**. The initial slope of the stress-strain curve from

の境界面で供試体を強制的にせん断させることで，せん断強度を求める試験機である。試験方法には，供試体の垂直変位が生じないように垂直応力を制御し，体積を一定に保ってせん断する定体積試験（図 8.14（b-①））と，垂直応力がせん断中一定値であるとする定圧試験（図（b-②））がある。

試験は，数段階の垂直荷重において，圧密を行った後，各垂直荷重においてせん断を行い，各垂直応力におけるせん断応力，せん断変位の関係を求める。強度定数は，定体積試験の場合，**図 8.15(a)**に示すように，各垂直応力における試験結果を基に応力経路を描き，この応力経路の包絡線を基に $\tau$ 軸の切片を粘着力 $c$，包絡線の傾きを内部摩擦角 $\phi$ とする。また，定圧試験の場合，垂直応力が一定であるので，図(b)に示すように，垂直応力とせん断強さをプロットし，この点を結んだ直線を基に $c$，$\phi$ を求める。

### （2）一軸圧縮試験

**図 8.16** に一軸圧縮試験機の概要を示す。円柱状の供試体を周りからの拘束のない状態で，軸方向力だけを与えて圧縮し，このときの圧縮応力を求める試験である。一軸圧縮試験により求まる最大の圧縮応力は，一軸圧縮強さ $q_u$ といわれる。この試験は，三軸圧縮試験の拘束圧が 0 の試験に相当するもので，

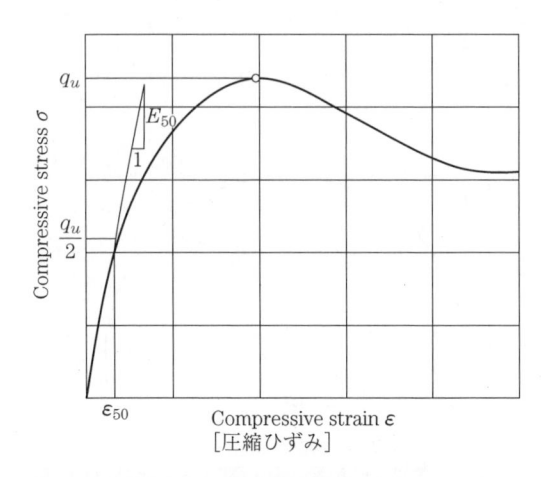

**Figure 8.17** Compressive stress-strain relationship from an unconfined compressive strength test
図 8.17 圧縮ひずみと圧縮応力の関係

the test results is called the "Modulus of deformation" and is given by the following equation:

$$E_{50} = \frac{\frac{q_u}{2}}{\varepsilon_{50}} \Big/ 10 \tag{8.22}$$

where

$\varepsilon_{50}$ : Compressive strain at $\frac{q_u}{2}$ in percent

The units of the modulus of deformation defined in Equation (8.22) are $N/m^2$ or Pa (Pascals).

The unconfined compressive strength test is performed without radial pressure i.e. ($\sigma_3 = 0$ kN/m²). Mohr's stress circle is plotted (**Figure 8.18**) using $q_u$ which is equivalent to $\sigma_1$. For clay soil, the failure envelope is flat ($\phi = 0°$), and therefore, the cohesion is equal to the radius of Mohr's stress circle. The cohesion is given by the following equation:

$$c = \frac{q_u}{2} \tag{8.23}$$

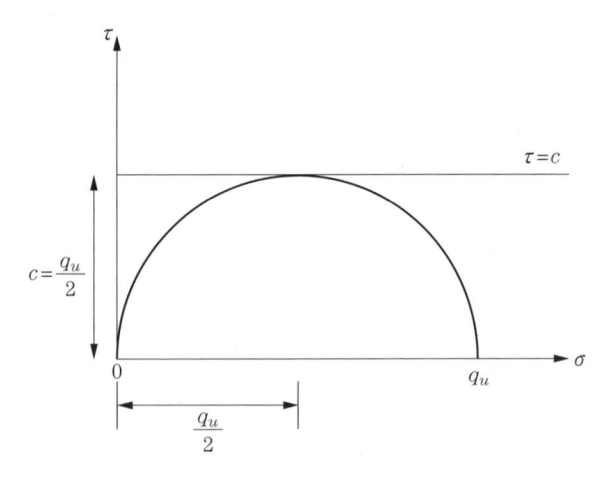

**Figure 8.18** Mohr's stress circle for unconfined compressive strength test. For clay soil, $\tau = c$
図 8.18　一軸圧縮試験におけるモールの応力円

試験は容易に実施できるが，拘束圧がないため砂のような自立できない供試体では試験が行えない。

図8.17は一軸圧縮試験結果の一例を示したものである。試験によって得られた応力-ひずみ曲線の初期の勾配を変形係数と呼び

$$E_{50} = \frac{\frac{q_u}{2}}{\varepsilon_{50}} \bigg/ 10 \qquad (8.22)$$

で求めることができる。

ここで，$\varepsilon_{50}$：$\frac{q_u}{2}$のときの圧縮ひずみ（％）である。

なお，式(8.22)で求まる変形係数の単位は$MN/m^2$である。

一軸圧縮試験は，拘束圧がなく$\sigma_3=0\,kN/m^2$であるので，$q_u$を$\sigma_1$としてモールの応力円を描くと図8.18になる。$\phi=0°$の粘土の場合，破壊包絡線は水

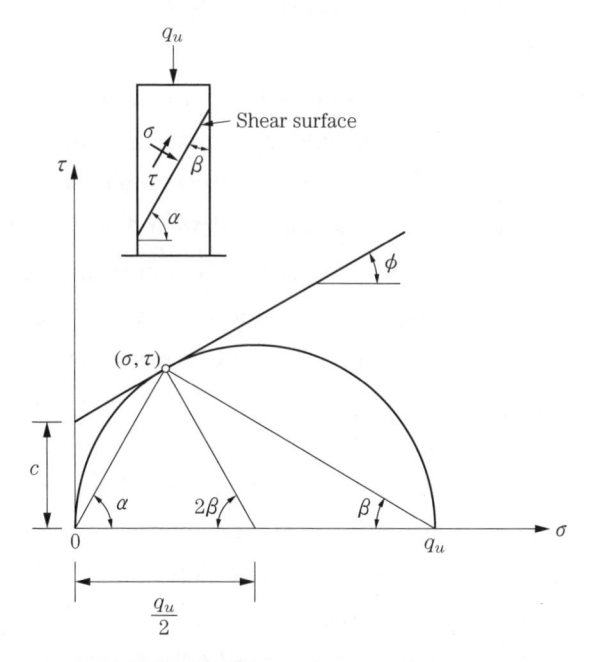

**Figure 8.19** The $(\sigma, \tau)$ on a shear surface from an unconfined compressive strength test
図8.19 一軸圧縮試験でのせん断面における $\sigma, \tau$

Accordingly, the cohesion is half of the unconfined compressive strength.

When a shear surface is formed at an angle $a°$ in a specimen tested in the laboratory, the stress on the shear surface can be determined by using the Mohr-Coulomb's failure criterion and Mohr's stress circle as shown in **Figure 8.19**. The strength of a disturbed-clay soil is reduced because of the breakdown of its structure. The sensitivity ratio $S_t$ is used to evaluate the degree of reduction in strength. The sensitivity ratio $S_t$ is defined as the ratio of unconfined compressive strength of undisturbed-soil $q_u$ to the unconfined compressive strength of disturbed-soil $q_{ur}$ and is given by:

$$S_t = \frac{q_u}{q_{ur}} \tag{8.24}$$

(3) **Triaxial Compressive Strength Test**

A sketch of a triaxial compressive testing apparatus is shown in **Figure 8.20**. Similar to the conditions in the ground, a cylindrical specimen is subjected to a normal stress while a radial stress is used to confine (restrict) the circumference of the sample. The compressive and/or shear strength of the sample is determined by making measurements for several radial pressures. The major and minor principal stresses for each radial stress condition are plotted on a Mohr's stress circle as shown in **Figure 8.21**. A failure envelope is drawn that connects the points that satisfy all the failure conditions for each stress circle. Then, the strength parameters $c$ and $\phi$ are determined from the failure envelope. In a triaxial compressive strength test, the consolidation conditions and drainage conditions can also be controlled and the shear strength can be determined as a function of consolidation and/or pore pressures.

平となり，粘着力はモールの応力円の半径に相当し

$$c = \frac{q_u}{2} \tag{8.23}$$

となり，粘着力は一軸圧縮強さの 1/2 で表される。

試験により供試体が角度 $α°$ の面でせん断面ができたとき，せん断面で発揮される応力 $(σ, τ)$ は，モール・クーロンの破壊基準を基に，図 8.19 に示すモールの応力円を描いて推定することができる。

粘土は練り返すことによって構造が壊れるため，強度が低下することがある。この強度が低下する程度を示す指標として鋭敏比 $S_t$ が用いられる。鋭敏比は，一軸圧縮試験により求まる乱さない土の一軸圧縮強さ $q_u$ と，練返しを行った土の一軸圧縮強さ $q_{ur}$ との比

$$S_t = \frac{q_u}{q_{ur}} \tag{8.24}$$

により求められる。

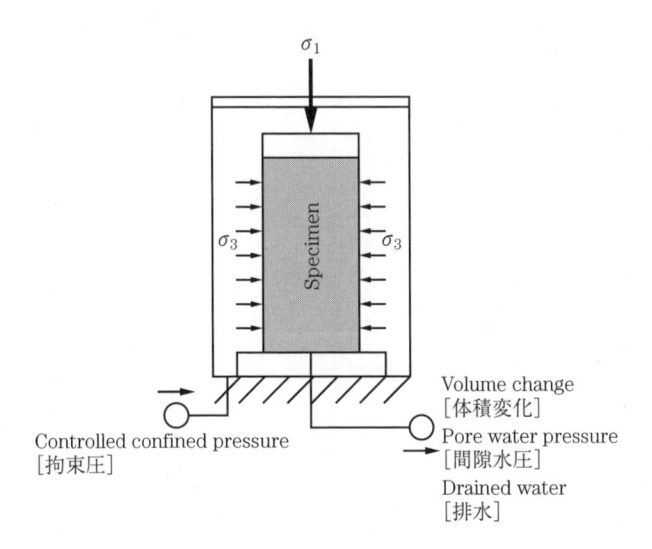

**Figure 8.20** A sketch of a triaxial compressive testing apparatus
図 8.20 三軸試験装置の概要

## Chapter 8 Shearing of Soil

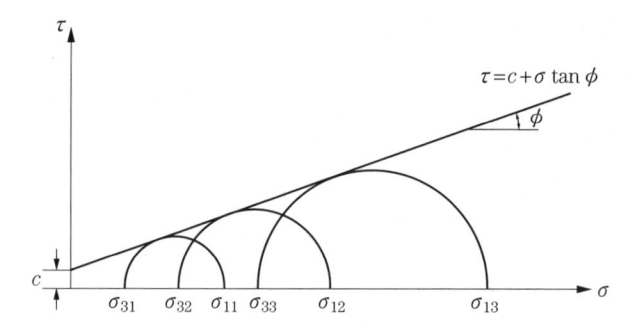

$$\tau = c + \sigma \tan \phi$$

**Figure 8.21** Mohr's stress circle for each radial stress from a triaxial compressive test
図 8.21 各段階の応力におけるモールの応力円

**Table 8.3** Test results
表 8.3 試験結果

| Compressive strain [%] | Compressive stress [kN/m²] | |
|---|---|---|
| | Undisturbed-soil [乱さない土] | Disturbed-soil [練返した土] |
| 0.00 | 0.0 | 0.0 |
| 0.25 | 7.0 | 0.5 |
| 0.50 | 14.0 | 1.0 |
| 0.75 | 21.0 | 1.5 |
| 1.00 | 28.0 | 2.0 |
| 1.25 | 35.0 | 2.5 |
| 1.50 | 40.0 | 3.0 |
| 1.75 | 43.0 | 3.5 |
| 2.00 | 46.0 | 4.0 |
| 2.25 | 48.0 | 4.5 |
| 2.50 | 49.0 | 5.0 |
| 2.75 | 50.0 | 5.5 |
| 3.00 | 50.0 | 6.0 |
| 3.25 | 48.0 | 6.5 |
| 3.50 | 45.0 | 7.0 |
| 3.75 | 42.0 | 7.5 |
| 4.00 | 40.0 | 8.0 |
| 4.50 | | 9.0 |
| 5.00 | | 10.0 |
| 6.00 | | 12.0 |
| 7.00 | | 14.0 |
| 8.00 | | 15.5 |
| 9.00 | | 17.0 |
| 10.0 | | 18.0 |
| 12.5 | | 19.0 |
| 15.0 | | 19.0 |

## （3）三軸圧縮試験

図 8.20 に三軸試験装置の概要を示す。円筒状の供試体を用い，地盤内のように周囲からの拘束（拘束圧）を受けた状態において，土の圧縮強さあるいはせん断強度を求める試験で，試験は数段階の拘束圧について行う。図 8.21 に示すように，強度定数 $c$, $\phi$ は，各段階の拘束圧における最大主応力，最小主応力を基にモールの応力円を描き，各応力円上で破壊条件を満足する点を連ねた破壊包絡線から決定する。三軸圧縮試験は，圧密条件，排水条件をコントロールできるため，各種条件を考慮したせん断強度を求めることが可能である。

**【例題 8.3】** 乱さない土と練り返した土を用いて一軸圧縮試験を行い，表 8.3 の試験結果を得た。このときの一軸圧縮強さを求めるとともに，変形係数および鋭敏比を求めよ。

**解答** 表 8.3 に示す結果を基に，乱さない土および練り返した土の圧縮ひずみ-圧縮応力の関係をグラフに示すと図 8.22 となる。この図を基に一軸圧縮強さを求めると，

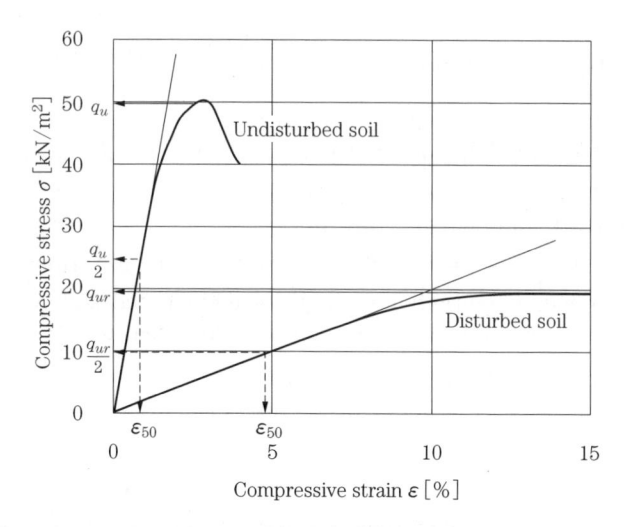

**Figure 8.22** Compressive stress–strain relationships based on data in Table 8.3 from the example
図 8.22 圧縮ひずみと圧縮応力の関係

# Chapter 8 Shearing of Soil

**〖Example 8.3〗** Results from unconfined compressive strength tests of disturbed-and undisturbed-soils are given in **Table 8.3**. Calculate the unconfined compressive strength, modulus of deformation and the sensitivity ratio.

**Solution** Based on the results given in Table 8.3, the compressive stress-strain relationships for the disturbed-and the undisturbed-soils are drawn as shown in **Figure 8.22**. From the graph, the compressive strengths are obtained as follows:

For undisturbed-soil: $q_u = 50 \text{ kN/m}^2$
For disturbed-soil: $q_{ur} = 19 \text{ kN/m}^2$

By using $E_{50} = \dfrac{\dfrac{q_u}{2}}{\varepsilon_{50}} \Big/ 10$, the modulus of deformation is calculated by finding $\varepsilon_{50}$ corresponding to $\dfrac{q_u}{2}$ in Figure 8.20.

The following results are obtained:

For undisturbed-soil: The compressive strain is ($\varepsilon_{50} = 0.89\,\%$) when $\dfrac{q_u}{2} = 25 \text{ kN/m}^2$

For disturbed-soil: The compressive strain is ($\varepsilon_{50} = 4.75\,\%$) when $\dfrac{q_u}{2} = 9.5 \text{ kN/m}^2$

Accordingly, the modulus of deformation of both soils are as follows:
For undisturbed-soil: $E_{50} = 2.8 \text{ MN/m}^2$
For disturbed-soil: $E_{50} = 0.2 \text{ MN/m}^2$

Sensitivity ratio $S_t$ is calculated using the following equation:

$$S_t = \frac{\text{Unconfined compressive strength of undisturbed-soil}}{\text{Unconfined compressive strength of disturbed-soil}} \frac{q_u}{q_u}$$

Therefore, $S_t = \dfrac{50}{19} = 2.63$

乱さない土：$q_u = 50\,\mathrm{kN/m^2}$

練り返した土：$q_{ur} = 19\,\mathrm{kN/m^2}$

となる。

変形係数 $E_{50}$，$E_{50} = \dfrac{\dfrac{q_u}{2}}{\varepsilon_{50}} \Big/ 10$ を用い，図 8.20 を基に $\dfrac{q_u}{2}$ のときの $\varepsilon_{50}$ を求めると

乱さない土：$\dfrac{q_u}{2} = 25\,\mathrm{kN/m^2}$ のときの圧縮ひずみ $\varepsilon_{50} = 0.89\,\%$

練り返した土：$\dfrac{q_{ur}}{2} = 9.5\,\mathrm{kN/m^2}$ のときの圧縮ひずみ $\varepsilon_{50} = 4.75\,\%$

より，両者の変形係数は

乱さない土：$E_{50} = 2.8\,\mathrm{MN/m^2}$

練り返した土：$E_{50} = 0.2\,\mathrm{MN/m^2}$

となる。

鋭敏比 $S_t$ は，

$$S_t = \frac{\text{乱さない土の一軸圧縮強さ } q_u}{\text{練返しを行った土の一軸圧縮強さ } q_{ur}}$$

により求められることより

$$S_t = \frac{50}{19} = 2.63$$

である。

---

## Jean Baptiste Joseph Fourier（フーリエ，1768〜1830）

フランスの数学者・物理学者。固体内での熱伝導に関する研究から熱伝導方程式を導き，これを解くためにフーリエ解析と呼ばれる理論を展開した。

フーリエは幼くして仕立屋の父を失い，地元の司教に預けられ，その司教は彼を陸軍幼年学校に入学させた。フーリエはここで数学に興味を示し勉強に没頭した。1789 年，論文を発表するためにパリへ向かい，そこでフランス革命に遭遇。革命後，新しく創設された高等教育機関に入学，才能を認められたフーリエはラグランジュなどのもとでエコール・ポリテクニークの助講師，のちに解析学の教授になった。土質力学でも，圧密の基礎方程式の解を求めるときなど，多くの場所でフーリエ級数が用いられている。

図12　フーリエ

## 8.6 Consolidated and Drained Conditions

The properties of a soil vary widely depending on the past and current consolidation and drainage conditions. Therefore, it is very important to study the consolidation and drainage conditions of a soil. First, we will consider consolidation for drained and undrained conditions. For undrained consolidation, volume changes cannot occur because of the presence of fluids (pore-water) in the voids (pores) and thus consolidation of the soil does not occur. This is similar to the unconsolidated condition for a constant void ratio $e$. On the other hand, for drained consolidation, pore-water can move freely and drain from the pores during consolidation causing the void ratio $e$ to decrease. This results in an increase in density as the void ratio $e$ decreases during consolidation. Therefore, the strength of a soil increases.

Next, we consider the shear strength of a soil for drained and undrained conditions when the soil is subjected to shear stress. For an applied shear stress to an undrained soil, the pore-water cannot drain and excess pore water pressure develops. Thus, the total stress must be used to determine the shear strength. On the other hand, for drained conditions when subjected to shear stresses, the pore-water moves freely and excess pore water pressure does not develop. Thus, the shear strength is determined using the effective stress.

As described above, the properties and shear strength of soil vary widely depending on the consolidation and drainage conditions. The soil properties and shear strength are needed for the design of safe structures in soil. Therefore it is important to perform shear tests on a soil based on the field conditions found at the site. There are three methods for performing shear tests using the triaxial compres-

## 8.6 圧密・排水条件

　土は，圧密・排水の条件によって力学特性が大きく異なるため，圧密と排水の条件を考えることが重要である。

　ここでまず，圧密時の排水の有無を考えてみる。圧密時に排水を許さない場合，体積が変化できないため圧密は生じず，間隙比 $e$ が一定の非圧密状態となる。一方，圧密時に排水を許した場合，間隙中の水が排水され圧密状態となり，間隙比 $e$ は小さくなる。供試体が圧密すると，間隙比 $e$ が小さくなるため密度は大きくなり，土のもっている強度は増加すると考えられる。

　つぎに，せん断時における排水の有無を考えてみる。せん断時に排水を許さない非排水条件では，せん断中に発生する間隙水圧が消散できず過剰間隙水圧の発生がみられるため，せん断強度は全応力に対する評価となる。一方，せん断時に排水を許す排水条件では，せん断中に間隙水圧が消散するため過剰間隙水圧の発生がなく，せん断強度は有効応力に対する評価となる。

---

**George Gabriel Stokes**（ストークス，1819～1903）

　アイルランドの数学者，物理学者。ケンブリッジ大学で学ぶ。流体力学，光学，数学などの分野で貢献した。王立協会会長。同じ頃，国会議員，Lucasian 教授になった。三つの地位を得た人はストークス以外にニュートンがいるが，同時期に三つの地位を得たのはストークスだけである。

　粘性流体のナビエ・ストークスの式，流体中を落下する粒子の速度を表すストークスの式，水面波のストークス波，粘度の単位ストークス，光ではストークス散乱，数学ではストークスの定理で知られる。

図13　ストークス

sive strength test for different consolidation and drainage conditions. The three methods are:, ① Unconsolidated and Undrained (UU) test, ② Consolidated and Undrained (CU, $\overline{CU}$) test and ③ Consolidated and Drained (CD) test.

### (1) Unconsolidated and Undrained (UU) Test

The unconsolidated and undrained (UU) test does not allow the pore water to drain before and during the shear test. In this test, the void ratio does not change because there is no consolidation. So, the shear strength remains constant. Performing a shear test under undrained conditions does not change the effective stress because the pore water pressure increases as the stress increases. Consequently, in UU tests, the followings parameters are constant, $e$, $\sigma'$ and $\tau$. The Mohr's stress circle moves to the right as the pore water pressure increases as shown in **Figure 8.23**. Therefore, the failure envelope is flat and the strength parameters are obtained as follows:

$$\phi_u = 0 \tag{8.25}$$

$$\phi_u = \frac{\sigma_1 - \sigma_3}{2} \tag{8.26}$$

In unconfined compressive strength tests, shear testing is performed on a clayey soil at relatively high speeds without consolidation. So, it can be considered as an UU test with zero radial pressure, i.e. $\sigma_3 = 0 \text{ N/m}^2$. For a clayey soil with ($\phi = 0°$), the equation for cohesion in terms of unconfined compressive strength $q_u$ and $\sigma_1$ is

$$c_u = \frac{\sigma_1}{2} = \frac{q_u}{2} \tag{8.27}$$

When a structure is rapidly constructed on a clayey soil, the ground cannot consolidate and is initially undrained. When designing for this type of rapid loading, it is necessary to use the unconsolidated and undrained shear strength parameters. This type of design problem is called the "short-term stability problem".

## 8.6 Consolidated and Drained Conditions

このように，構造物の設計や安定性の評価に用いるせん断強度は，圧密，排水条件によって大きく特性が異なるため，せん断試験を実施する場合，現地に対応した条件を考慮して行うことが重要である。

三軸圧縮試験などのせん断試験を実施する場合，圧密・排水を考慮して，①非圧密・非排水（UU）条件，②圧密・非排水（CU，$\overline{\mathrm{CU}}$）条件，③圧密・排水（CD）条件の三つの試験法が存在する。

### （1） 非圧密・非排水（UU）試験

UU試験は，せん断前およびせん断中に供試体から排水を許さない試験方法である。UU試験では，圧密が行われないため間隙比の変化がなく，せん断強度は一定となる。また，非排水状態でせん断を行うと，増加した応力はすべて間隙水圧で受けもたれるため，有効応力は変化しない。したがって，UU試験は，$\sigma$ の増加に対して，$e$：一定，$\sigma'$：一定，$\tau$：一定となり，モールの応力円は図 8.23 に示すように間隙水圧の増加分だけ移動するのみで，破壊包絡線は水平な直線となり，得られる強度定数は

$$\phi_u = 0 \tag{8.25}$$

$$\phi_u = \frac{\sigma_1 - \sigma_3}{2} \tag{8.26}$$

となる。

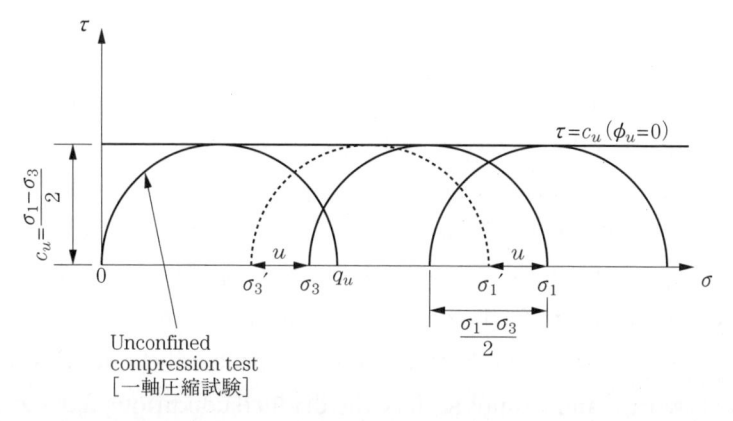

**Figure 8.23** Mohr's stress circle for unconsolidated and undrained (UU) tests
図 8.23 非圧密・非排水試験におけるモールの応力円

76　Chapter 8　Shearing of Soil

( 2 )　**Consolidated and Undrained (CU, $\overline{\text{CU}}$) Tests**

Consolidated and undrained (CU, $\overline{\text{CU}}$) tests allow drainage and consolidation prior to shearing but no drainage is allowed from the specimen during shearing. In CU tests, a decrease in void ratio from consolidation occurs during shearing and causes an increase in the shear strength of a soil. Therefore, in CU tests, as $\sigma$ increases, $e$ decreases, $\sigma'$ increases, $\tau$ increases and the Mohr's circle behaves as shown by the solid lines in **Figure 8.24**. When the effective stress is determined by measuring excess pore-water pressure and the total stress during shearing of a saturated specimen in a CU test, it is called a $\overline{\text{CU}}$ test.

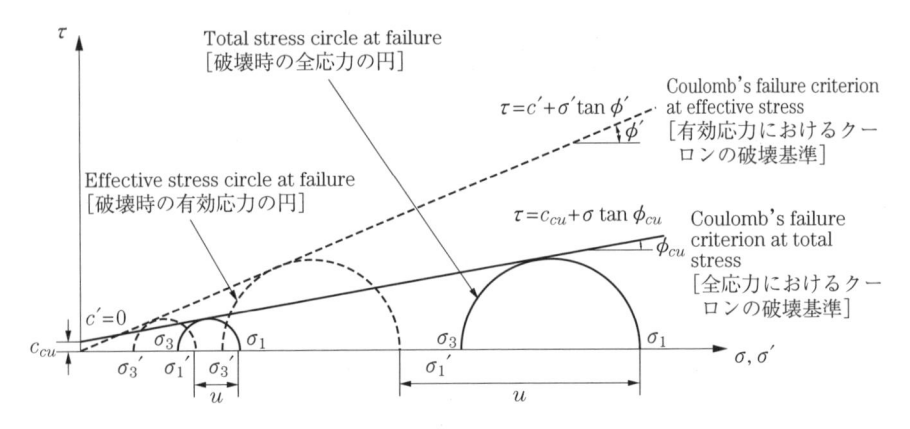

**Figure 8.24**　Mohr's stress circle for consolidated and undrained tests
　　図 8.24　圧密・非排水試験におけるモールの応力円

For practical construction, the shear strength of a sand or gravel can be measured by using the CD tests because sand and gravel have large coefficients of permeability that enables the soil to drain easily, and thus satisfy the drained conditions during shear. However, a clayey soil usually has a low value of permeability and drains at a very slow rate and cannot satisfy the drained conditions during shear. Therefore, in case of measuring effective stress in a clayey soil, $\overline{\text{CU}}$

一軸圧縮試験は，圧密を行わず比較的速い速度で粘性土のせん断を行うため，拘束圧 $\sigma_3=0\,\text{kN/m}^2$ の UU 試験とみなすことができる。このため，$\phi=0°$ の粘性土の場合，一軸圧縮強さ $q_u$ を $\sigma_1$ と考えると

$$c_u = \frac{\sigma_1}{2} = \frac{q_u}{2} \tag{8.27}$$

で示される。

実際の工事などにおいて，粘土地盤上に比較的短期間に構造物を構築する場合，施工後初期の段階では，地盤は圧密・排水することができない。このような急速載荷の設計においては，非圧密・非排水せん断強度を用いることが必要となる。このような設計問題を短期安定問題と呼ぶ。

（2）圧密・非排水（CU, $\overline{\text{CU}}$）試験

CU 試験（$\overline{\text{CU}}$ 試験）は，せん断前に排水を許すことで圧密させるものの，せん断中には供試体からの排水を許さない試験方法である。CU 試験では，せん断前に圧密を行うことで間隙比は小さくなり，せん断強度は増加する。したがって，CU 試験では，$\sigma$ の増加に対して，$e$：減少，$\sigma'$：増加，$\tau$：増加となり，モールの応力円は図 8.24 に示す実線のようになる。

ところで，供試体が飽和した CU 試験において，せん断中に全応力と併せて過剰間隙水圧の測定を行うことによって有効応力を求めることが可能となり，この試験を $\overline{\text{CU}}$ 試験と呼ぶ。

有効応力を求める場合，砂・礫などの透水係数が大きい材料では，排水が容易であるため，せん断中の排水条件を満足でき，CD 試験において有効応力を求めることは可能である。しかし，粘性土のように透水係数が小さい材料では，せん断中の排水条件を満足するため透水係数に見合った極端に遅いせん断スピードを採用する必要があり，現実的ではない。このため，粘性土の有効応力を求める場合，$\overline{\text{CU}}$ 試験を採用する必要がある。

圧密・非排水せん断試験によって得られる CU 試験のせん断強度は

$$\tau = c_{cu} + \sigma \tan\phi_{cu} \tag{8.28}$$

で示され，$c_{cu}$，$\phi_{cu}$ は全応力における粘着力，内部摩擦角である。

tests must be used.

The shear strength from consolidated and undrained (CU) tests is determined as follows:

$$\tau = c_{cu} + \sigma \tan \phi_{cu} \tag{8.28}$$

where, $c_{cu}$ and $\phi_{cu}$ are the cohesion and the angle of internal friction, respectively, based on total stress.

The shear strength for $(\overline{CU})$ tests is:

$$\tau = c' + \sigma' \tan \phi' = c' + (\sigma - u) \tan \phi' \tag{8.29}$$

where, $c'$ and $\phi'$ are the cohesion and the angle of internal friction, respectively, using effective stress ($\sigma' = \sigma - u$).

The Mohr's stress circle for effective stress shifts to left by an amount equal to the excess pore-water pressure $u$ from the Mohr's stress circles for total stress as shown by the dashed lines in Figure 8.24.

(3) **Consolidated and Drained (CD) Tests**

In consolidated and drained (CD) tests, consolidation is carried out before shearing and the water is gradually drained during shearing. Because the void ratio in the CD test decreases due to consolidation, the shear strength increases. In this test, excess pore-water pressure does not developed because drainage is allowed during shear. This test permits volume change during shear. In regards to the increase of ($\sigma = \sigma'$ ($u=0$)) in a CD test, $e$ decreases, $\sigma'$ increases, $\tau$ increases and the Mohr's circle behaves as shown in **Figure 8.25**. The effective stress can be measured directly from the consolidated and drained test results because the excess pore-water pressure becomes zero ($u=0$) which satisfies the drained condition.

The shear strength for a CD test is

$$\tau_f = c_d + \sigma' \tan \phi_d$$

where, $c_d$ and $\phi_d$ are the cohesion and the angle of internal friction, respectively, using effective stress.

また，$\overline{\mathrm{CU}}$ 試験のせん断強度は

$$\tau = c' + \sigma' \tan\phi' = c' + (\sigma - u)\tan\phi' \tag{8.29}$$

で示され，$c'$, $\phi'$ は有効応力（$\sigma' = \sigma - u$）における粘着力，内部摩擦角である。なお，有効応力におけるモールの応力円は，全応力のモールの応力円が過剰間隙水圧 $u$ の分だけ左へ移動した図 8.24 の破線のようになる。

### （3） 圧密・排水（CD）試験

CD 試験は，せん断前に圧密をさせた後，排水を許しながらせん断を行う試験方法である。CD 試験では，圧密により間隙比は小さくなり，せん断強度は増加する。また，せん断中に排水を許すことでせん断時の過剰間隙水圧の発生がなく，せん断時の体積変化を許す条件でもある。CD 試験では，$\sigma = \sigma'$（$u=0$）の増加に対して，$e$：減少，$\sigma'$：増加，$\tau$：増加となり，モールの応力円は図 8.25 に示すようになる。

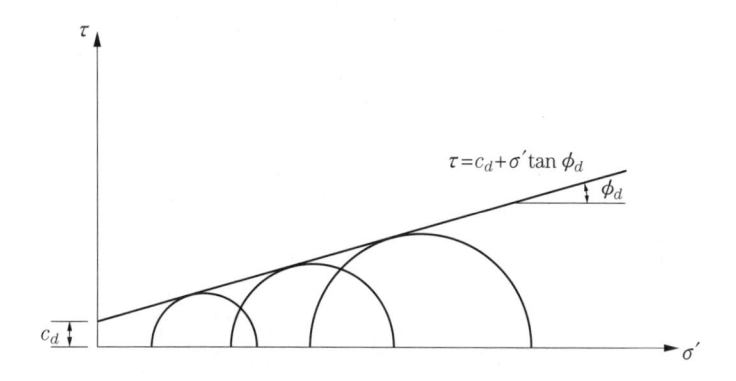

**Figure 8.25** Mohr's stress circle for consolidated and drained conditions
図 8.25  圧密・排水試験におけるモールの応力円

圧密・排水せん断試験では，排水条件を満足するため，過剰間隙水圧 $u=0$ となることから有効応力を直接求めることができる。

CD 試験のせん断強度は

$$\tau_f = c_d + \sigma' \tan\phi_d$$

で示され，$c_d$, $\phi_d$ は有効応力に相当する粘着力，内部摩擦角である。

For practical construction, the CD tests provide strength parameters for cases where "stability problems of a sandy soil that satisfies the drained conditions because it has a high permeability" or for cases where "long-term consolidation problems of a clayey soil occur because of a slow rate of construction". In these cases, the strength parameters based on consolidated and drained conditions need to be considered. Design problems for cases with a slow loading rate and that consider the effect of consolidation are called "long-term stability problems".

【Example 8.4】 Shear tests on a sand under $\overline{CU}$ conditions were performed with a radial stress of ($\sigma_3 = 50$ kN/m²). The results that

Table 8.4 Test results
表 8.4 実験結果

| Compressive strain [%] | Difference in principal stresses [主応力差] $\sigma_1 - \sigma_3$ [kN/m²] | Pore-water pressure [間隙水圧] $u$ [kN/m²] |
|---|---|---|
| 0.00 | 0.0 | 0.0 |
| 0.25 | 7.0 | 3.0 |
| 0.50 | 14.0 | 6.0 |
| 0.75 | 21.0 | 8.0 |
| 1.00 | 28.0 | 9.0 |
| 1.25 | 35.0 | 10.0 |
| 1.50 | 40.0 | 10.0 |
| 1.75 | 43.0 | 10.0 |
| 2.00 | 46.0 | 10.0 |
| 2.25 | 48.0 | 10.0 |
| 2.50 | 49.0 | 10.0 |
| 2.75 | 50.0 | 10.0 |
| 3.00 | 50.0 | 10.0 |
| 3.25 | 48.0 | 10.0 |
| 3.50 | 47.0 | 10.0 |
| 3.75 | 46.0 | 10.0 |
| 4.00 | 45.0 | 10.0 |

**Table 8.5** Results of total stress and effective stress
**表 8.5** 全応力と有効応力の計算結果

| Compressive strain [%] | Difference in principal stresses $\sigma_1-\sigma_3$ [kN/m²] | Pore-water pressure [kN/m²] | $\sigma_1$ | $\sigma_1'$ | $\sigma_3'$ | $p'$ $\frac{\sigma_1'+\sigma_3'}{2}$ | $\frac{\sigma_1'-\sigma_3'}{2}$ | $p$ $\frac{\sigma_1+\sigma_3}{2}$ | $q$ $\frac{\sigma_1-\sigma_3}{2}$ |
|---|---|---|---|---|---|---|---|---|---|
| 0.00 | 0.0 | 0.0 | 50.0 | 50.0 | 50.0 | 50.0 | 0.00 | 50.0 | 0.0 |
| 0.25 | 7.0 | 3.0 | 57.0 | 54.0 | 47.0 | 50.5 | 3.50 | 53.5 | 3.5 |
| 0.50 | 14.0 | 6.0 | 64.0 | 58.0 | 44.0 | 51.0 | 7.00 | 57.0 | 7.0 |
| 0.75 | 21.0 | 8.0 | 71.0 | 63.0 | 42.0 | 52.5 | 10.5 | 60.5 | 10.5 |
| 1.00 | 28.0 | 9.0 | 78.0 | 69.0 | 41.0 | 55.0 | 14.0 | 64.0 | 14.0 |
| 1.25 | 35.0 | 10.0 | 85.0 | 75.0 | 40.0 | 57.5 | 17.5 | 67.5 | 17.5 |
| 1.50 | 40.0 | 10.0 | 90.0 | 80.0 | 40.0 | 60.0 | 20.0 | 70.0 | 20.0 |
| 1.75 | 43.0 | 10.0 | 93.0 | 83.0 | 40.0 | 61.5 | 21.5 | 71.5 | 21.5 |
| 2.00 | 46.0 | 10.0 | 96.0 | 86.0 | 40.0 | 63.0 | 23.0 | 73.0 | 23.0 |
| 2.25 | 48.0 | 10.0 | 98.0 | 88.0 | 40.0 | 64.0 | 24.0 | 74.0 | 24.0 |
| 2.50 | 49.0 | 10.0 | 99.0 | 89.0 | 40.0 | 64.5 | 24.5 | 74.5 | 24.5 |
| 2.75 | 50.0 | 10.0 | 100.0 | 90.0 | 40.0 | 65.0 | 25.0 | 75.0 | 25.0 |
| 3.00 | 50.0 | 10.0 | 100.0 | 90.0 | 40.0 | 65.0 | 25.0 | 75.0 | 25.0 |
| 3.25 | 48.0 | 10.0 | 98.0 | 88.0 | 40.0 | 64.0 | 24.0 | 74.0 | 24.0 |
| 3.50 | 47.0 | 10.0 | 97.0 | 87.0 | 40.0 | 63.5 | 23.5 | 73.5 | 23.5 |
| 3.75 | 46.0 | 10.0 | 96.0 | 86.0 | 40.0 | 63.0 | 23.0 | 73.0 | 23.0 |
| 4.00 | 45.0 | 10.0 | 95.0 | 85.0 | 40.0 | 62.5 | 22.5 | 72.5 | 22.5 |

were obtained are shown in **Table 8.4**. Calculate the angle of internal friction using the total stress and the effective stress. Also, show the stress paths for both total stress and effective stress.

**Solution**  Based on the data given in **Table 8.5**, Mohr's stress circles for total stress $\sigma$ and effective stress $\sigma'$ are drawn as shown in **Figure 8.26**. The angle of internal friction is calculated from the failure envelope that connects the Mohr's stress circle to the origin because the cohesion ($c=0$ N/m²) is zero for a sand. By using $\sin\phi = \dfrac{\sigma_1 - \sigma_3}{\sigma_1 + \sigma_3}$,

For total stress, $\sigma_1 = 100$ kN/m² and $\sigma_3 = 50$ kN/m². Hence, the angle of internal friction : $\phi = 19.5°$

For effective stress, $\sigma_1' = 90$ kN/m² and $\sigma_3' = 40$ kN/m². Hence, the angle of internal friction : $\phi' = 22.6°$

Based on results given in Equations (8.20), (8.21) and on $p = \dfrac{\sigma_1 + \sigma_3}{2}$, $q = \dfrac{\sigma_1 - \sigma_3}{2}$ and $p' = \dfrac{\sigma_1' + \sigma_3'}{2}$, the stress paths for total stress $\sigma$ and effective stress $\sigma'$ are drawn as shown in **Figure 8.27**.

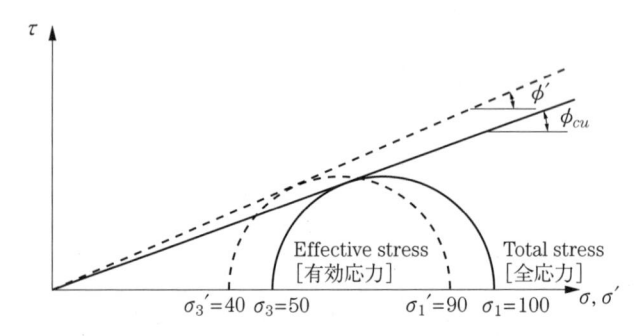

**Figure 8.26**  Mohr's stress circle for total stress and effective stress
図 8.26  全応力と有効応力におけるモールの応力円

## 8.6 Consolidated and Drained Conditions

実際の工事において，排水条件を満足できる透水性の良い砂質土の安定問題や，盛土などの緩速施工により盛土下の粘土地盤の圧密を考える長期的な問題の場合には，圧密・排水せん断強度を用いることが必要であり，このような圧密の影響を考慮する「緩速載荷」の設計問題を長期安定問題と呼ぶ．

**【例題 8.4】** 拘束圧 $\sigma_3=50\,\mathrm{kN/m^2}$ において $\overline{\mathrm{CU}}$ 条件で砂のせん断試験を行い，**表**8.4 の結果を得た．このとき，全応力および有効応力における内部摩擦角を求めよ．また，全応力，有効応力における応力経路を示せ．

**解答** **表**8.5 に示す全応力および有効応力における $\sigma$，$\sigma'$ を基に，モールの応力円を描くと**図**8.26 となる．砂であるので粘着力 $c=0\,\mathrm{kN/m^2}$ として，原点を通りモールの応力円に接する破壊包絡線から内部摩擦角を求めると，$\sin\phi=\dfrac{\sigma_1-\sigma_3}{\sigma_1+\sigma_3}$ の関係から，
全応力では $\sigma_1=100\,\mathrm{kN/m^2}$，$\sigma_3=50\,\mathrm{kN/m^2}$ より，内部摩擦角：$\phi=19.5°$
有効応力では $\sigma_1'=90\,\mathrm{kN/m^2}$，$\sigma_3'=40\,\mathrm{kN/m^2}$，内部摩擦角：$\phi'=22.6°$
となる．

式(8.20)，(8.21)を参考に $p=\dfrac{\sigma_1+\sigma_3}{2}$，$q=\dfrac{\sigma_1-\sigma_3}{2}$，$p'=\dfrac{\sigma_1'+\sigma_3'}{2}$ より，全応力および有効応力における応力経路を描くと，それぞれ**図**8.27 となる．

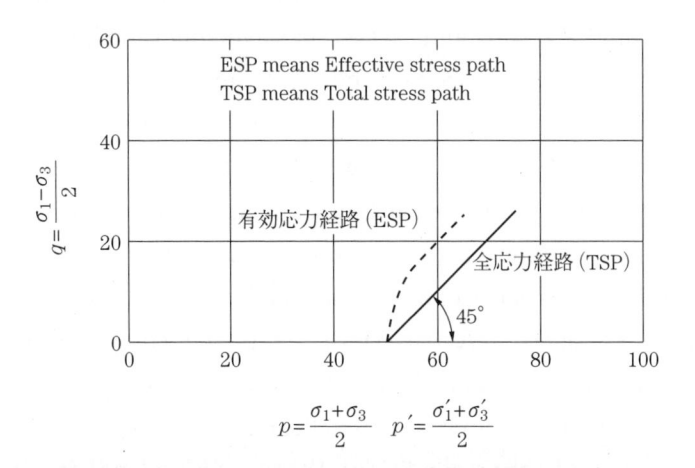

**Figure** 8.27　Stress paths for total stress and effective stress
**図** 8.27　全応力と有効応力の応力経路

# Chapter 9  Earth Pressure

## 9.1  Types of Earth Pressure

In the ground, the earth exerts pressure on a body in a manner similar to water pressure when a body is underwater. Water pressure is the same in all the directions (hydrostatic pressure) when it is at rest. However, unlike water pressure, the earth pressure is not the same in all directions because soil has shear strength. Thus, it is important to understand how earth pressure changes in a ground.

Let's consider an earth pressure acting on a vertical wall as shown in **Figure 9.1**. Three earth pressure conditions are shown: ① the "at rest condition", when a wall does not move, ② the "active condition", when a wall moves forward loosening or uncompressing the soil behind the wall and ③ the "passive condition", when a wall moves backward and pushes on the soil behind the wall causing it to compress.

① Condition 1 (At rest condition): When the wall is in the at rest condition, the earth pressure in the vertical direction at depth $z$ is ($\sigma_v = \gamma_t\, z$). The earth pressure in the horizontal (or lateral) direction is proportional to the vertical earth pressure and is expressed as ($\sigma_h = K_0\, \sigma_v$). Here, $K_0$ is the coefficient of earth pressure at rest. Generally, the horizontal earth pressure is half of the vertical earth pressure ($K_0 = 0.5$).

② Condition 2 (Active condition): When a wall moves forward, the earth pressure in the vertical direction at depth $z$ is same as the

# 9 章 土　　　　圧

## 9.1 土圧の種類

水中で水圧が作用するのと同様，地盤内では土圧が作用する。ところで，水圧は静水状態ではすべての方向で等しい圧力を示すが，土圧は土がせん断抵抗を有しているため，水圧と異なりすべての方向で同じ圧力を示さない。このため，地盤内の土圧がどのように変化するかを考えることが必要である。

図 9.1 に示すような垂直な壁に作用する土圧を考えてみる。このとき，①

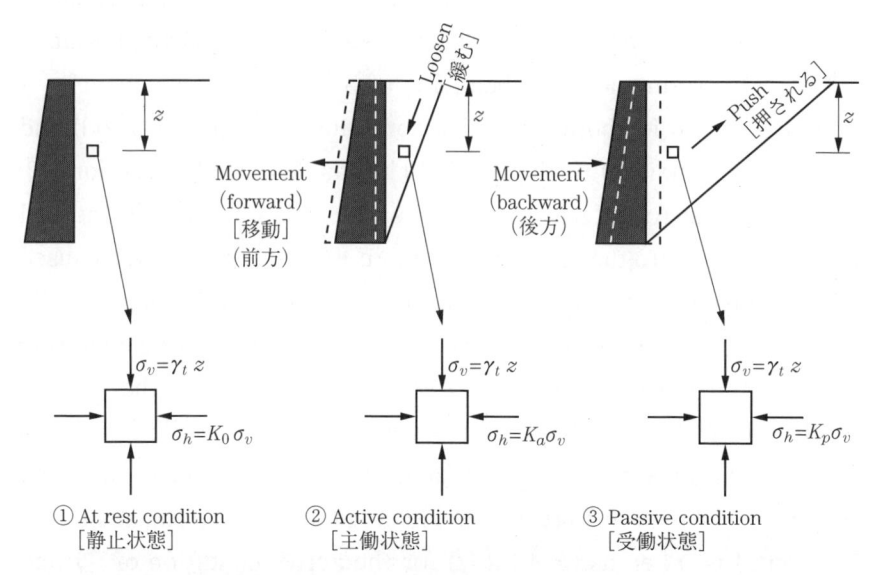

**Figure 9.1** Various earth pressures acting on a vertical wall in the ground
図 9.1　垂直な壁に作用する種々の土圧

"at rest earth pressure" i.e. ($\sigma_v = \gamma_t\, z$). However, the horizontal (or lateral) earth pressure is smaller than the "at rest earth pressure" because of the uncompressing or loosening of the soil behind the wall. The earth pressure at this condition is called an "active earth pressure". The relationship between the vertical earth pressure and the horizontal earth pressure for the active condition is given by ($\sigma_h = K_a \sigma_v$). Here, $K_a$ is the "coefficient of active earth pressure".

③ Condition 3 (Passive condition): When a wall moves backward, the earth pressure in the vertical direction at depth $z$ is same as the "at rest earth pressure" i.e. ($\sigma_v = \gamma_t\, z$). However, the horizontal earth pressure is larger than the "at rest earth pressure" because of the compression of the soil behind the wall. The earth pressure at this condition is known as the "passive earth pressure". The relationship between the vertical earth pressure and the horizontal earth pressure for the passive condition is given by ($\sigma_h = K_p\, \sigma_v$). Here $K_p$ is the "coefficient of passive earth pressure".

In **Figure 9.2**, the behavior of a soil after movement of a wall are shown for both the active condition (in Figure 9.2 (a)) and the passive condition (in Figure 9.2 (b)) as described in sections ② and ③ above. A slip surface (shear surface) is observed in the soil behind the wall (to the right of the wall in Figure 9.2) that was caused by the movement of the wall. The soil blocks moved along the slip surface. The slip surface is steep for the case of the active condition of ②, while the slip surface is gentle for the case of the passive condition of ③. In both cases, the slip surface is curvilinear. However, it is assumed to be linear to facilitate the calculation of the earth pressure as shown in **Figure 9.3**. The angle of the slip surface to the horizontal is taken as ($\pi/4 + \phi/2$) for the active condition of ② and as ($\pi/4 - \phi/2$) for the passive condition of ③.

## 9.1 Types of Earth Pressure

壁が静止している状態，②壁が前方に移動し背面が緩む状態，③壁が後方に移動し背面が押される状態の，三つの状態を考えてみる。

①の状態（静止状態）：壁が静止している状態では，地盤深さ $z$ の位置における垂直方向の土圧は $\sigma_v = \gamma_t z$ である。水平方向の土圧は，垂直土圧に比例すると考えて，$\sigma_h = K_0 \sigma_v$ で示される。ここで，$K_0$ は静止土圧係数と呼ばれる係数で，一般に静止状態では，水平土圧は垂直土圧の半分程度と考え，$K_0 = 0.5$ 程度である。

②の状態（主働状態）：壁が前方に移動する場合，地盤深さ $z$ の位置での鉛

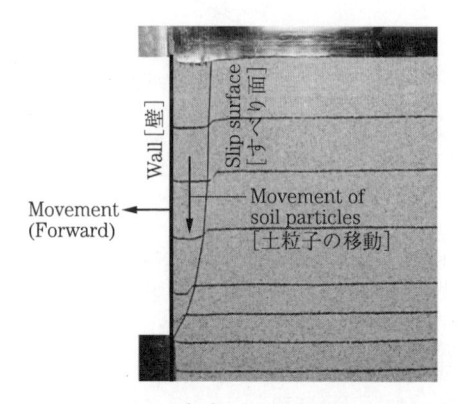

(a) Active condition

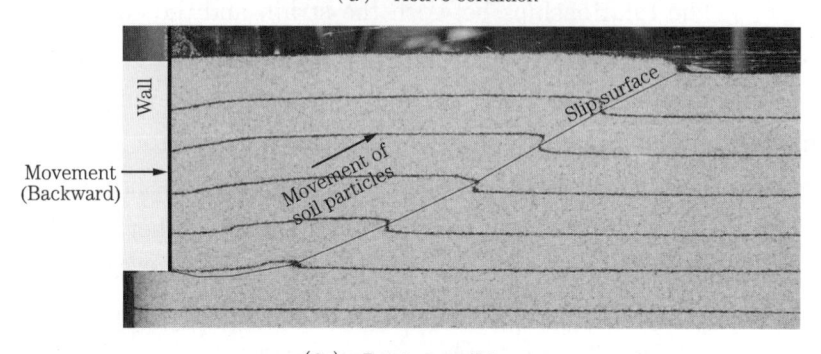

(b) Passive condition

**Figure 9.2** Photographs of soil subjected to the active condition (a) and the passive condition (b)
図 9.2 主働状態と受働状態

## Chapter 9  Earth Pressure

**Figure 9.3** Slip surfaces for active and passive conditions
図 9.3  主働状態と受働状態のすべり面の仮定

(a) Active condition — Movement (Forward), $\dfrac{\pi}{4}+\dfrac{\phi}{2}$

(b) Passive condition — Movement (Backward), $\dfrac{\pi}{4}-\dfrac{\phi}{2}$

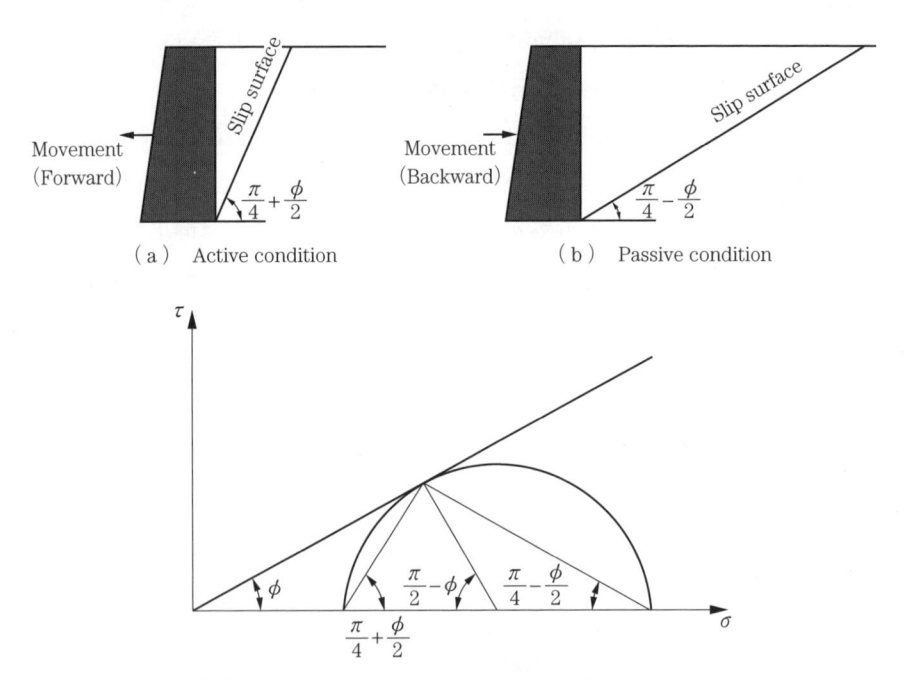

(c) The angle of the slip surface on Mohr's stress circle
[モールの応力円におけるすべり面の角度]

Angles shown: $\phi$, $\dfrac{\pi}{2}-\phi$, $\dfrac{\pi}{4}-\dfrac{\phi}{2}$, $\dfrac{\pi}{4}+\dfrac{\phi}{2}$

The relationships between the strain and the coefficient of earth pressure caused by the movement of a wall are shown in **Figure 9.4**. The coefficient of earth pressure for the "active condition" is smaller in value than the coefficient of earth pressure for the "at rest condition". Often, the strain for the "active (earth pressure) state" is taken to be approximately $-2 \sim -4\%$. On the other hand, the coefficient of earth pressure for the "passive condition" is larger in value than the coefficient of earth pressure for the "at rest condition". The strain for "passive (earth pressure) state" is greater than 15 %.

直方向の土圧は静止状態の場合と同じ $\sigma_v = \gamma_t z$ であるが，水平方向の土圧は背面が緩むため静止状態に比べ小さくなる。このときの土圧を主働土圧と呼び，主働状態における垂直土圧と水平土圧との関係は $\sigma_h = K_a \sigma_v$ で表され，$K_a$ を主働土圧係数と呼ぶ。

③の状態（受働状態）：壁が後方に移動する場合，地盤深さ $z$ の位置での垂直方向の土圧は静止状態の場合と同じ $\sigma_v = \gamma_t z$ であるが，水平方向の土圧は静止状態に比べ背面が押されるため大きくなる。このときの土圧を受働土圧と呼び，受働状態における鉛直土圧と水平土圧の関係は，$\sigma_h = K_p \sigma_v$ で表され，$K_p$ を受働土圧係数と呼ぶ。

図 9.2 は，上記 ②，③ の主働状態，受働状態において，壁の移動により背面の状況がどのようになるかを観察したものである。壁の移動によって背面の地盤にすべり面（せん断面）が現れ，このすべり面に沿って土塊が移動する。すべり面は，② の主働状態では傾きが急であるのに対し，③ の受働状態では緩くなる。

図 9.2 のように実際のすべり面の形状は曲線を示すが，土圧を求める場合，計算を簡便にするため図 9.3 のように，水平面に対するすべり面の角度を，上

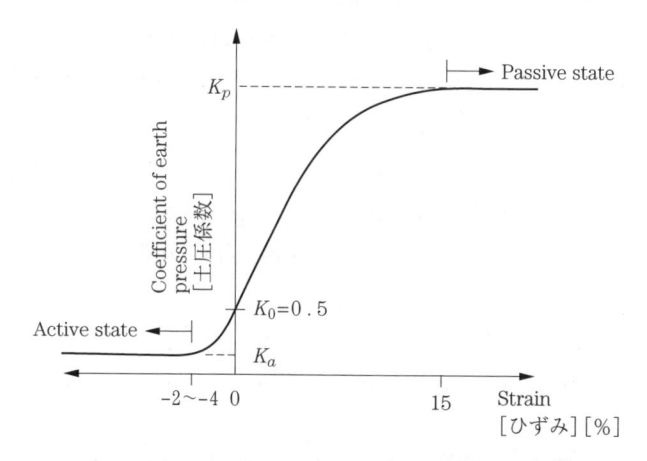

**Figure 9.4** Relationship between the strain and the coefficient of earth pressure caused by the movement of a wall
図 9.4 壁の移動によるひずみ量と土圧係数の関係

## 9.2 Rankine's Earth Pressure

In 1857, Rankine considered the case of a soil mass that exerts an earth pressure on the back of a wall with a vertical stress $\sigma_v$ and developed the relationship between the vertical pressure $\sigma_v$ and the horizontal pressure $\sigma_h$. For the active earth pressure condition, when a wall moves forward loosening the soil behind the wall, the relationship is ($\sigma_v > \sigma_h$). In this case, the magnitude of the horizontal earth pressure that causes compressive shear failure is taken as "an active earth pressure $\sigma_{ha}$".

On the other hand, for the passive condition, when a wall moves backward pushing on the soil behind the wall, the relationship between vertical pressure $\sigma_v$ and horizontal pressure $\sigma_h$ is ($\sigma_v < \sigma_h$). In this case, the magnitude of the horizontal earth pressure that causes expansion shear failure is taken as "a passive earth pressure $\sigma_{hp}$" (**Figure 9.5**). Rankine's earth pressure theory assumed a vertical wall and a flat horizontal ground behind the wall. The active and passive earth pressures are calculated assuming principal planes without any shear stresses on the horizontal and vertical surfaces inside the ground (Figure 9.5).

The vertical stress $\sigma_v$, and the horizontal stresses ($\sigma_{ha}$ and $\sigma_{hp}$) for active and passive conditions are expressed in terms of principal stresses ($\sigma_1$ and $\sigma_3$). Therefore, the Coulomb's failure criterion for principal stresses is

$$\sigma_1(1-\sin\phi) = \sigma_3(1+\sin\phi) + 2c\cos\phi \tag{9.1}$$

From Equation (9.1), the active and passive earth pressures can be calculated.

As described above, because ($\sigma_v > \sigma_{ha}$) for the active condition, $\sigma_v$ is the major principal stress $\sigma_1$, and $\sigma_{ha}$ is the minor principal

記②の主働状態では $\pi/4+\phi/2$,③の受働状態では $\pi/4-\phi/2$ とした直線として仮定する。

図 9.4 に,壁の移動によるひずみ量と土圧係数との関係を示す。主働土圧係数は静止土圧係数より小さく,主働土圧状態のひずみは（主働方向で）2〜4％程度といわれている。一方,受働土圧係数は静止土圧係数より大きく,受働土圧状態のひずみは（受働方向で）15％以上といわれている。

## 9.2 ランキン土圧

ランキンは,壁の背面地盤の垂直応力が $\sigma_v$ のとき,壁が前方に移動し背面が緩む主働状態では,垂直土圧 $\sigma_v$ と水平土圧 $\sigma_h$ の関係が $\sigma_v > \sigma_h$ となると考え,圧縮せん断破壊における水平土圧の大きさを主働土圧 $\sigma_{ha}$ とした。一方,

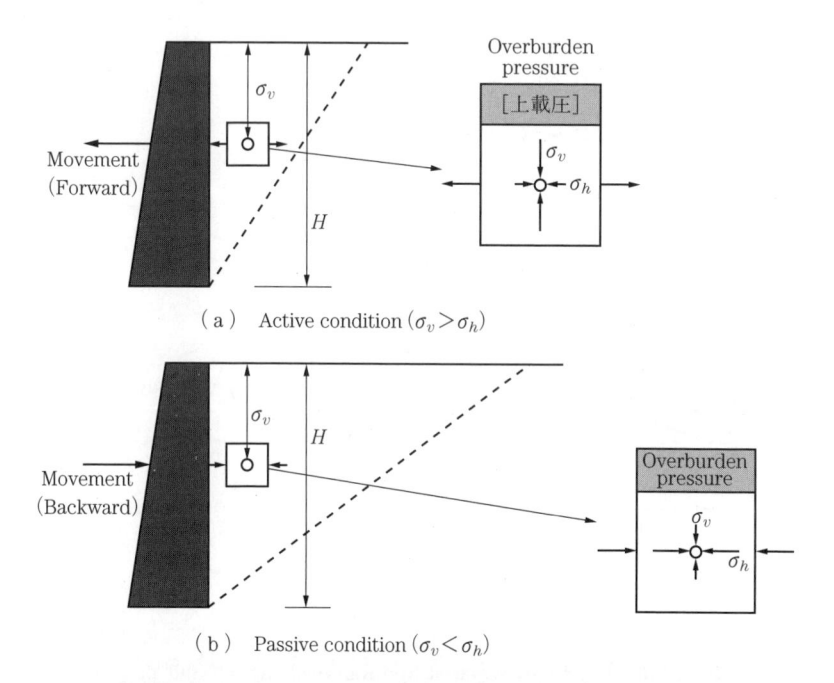

**Figure 9.5** Assumptions for Rankine's earth pressure theory
図 9.5 ランキンの土圧理論の仮定

stress $\sigma_3$. For the passive condition, because ($\sigma_v < \sigma_{hp}$), $\sigma_{hp}$ is the major principal stress $\sigma_1$, and $\sigma_v$ is the minor principal stress $\sigma_3$. Using the major and minor principal stresses in Equation (9.1) and drawing Mohr's stress circle (**Figure 9.6**), the active $\sigma_{ha}$ and passive $\sigma_{hp}$ earth pressures are calculated as follows,

$$\sigma_{ha} = \frac{1-\sin\phi}{1+\sin\phi}\sigma_v - 2c\frac{\cos\phi}{1+\sin\phi} \tag{9.2}$$

$$\sigma_{hp} = \frac{1+\sin\phi}{1-\sin\phi}\sigma_v + 2c\frac{\cos\phi}{1-\sin\phi} \tag{9.3}$$

where,

$$\frac{1\mp\sin\phi}{1\pm\sin\phi} = \tan^2\left(\frac{\pi}{4} \mp \frac{\phi}{2}\right) \tag{9.4}$$

$$\frac{\mp\cos\phi}{1\pm\sin\phi} = \tan\left(\frac{\pi}{4} \mp \frac{\phi}{2}\right) \tag{9.5}$$

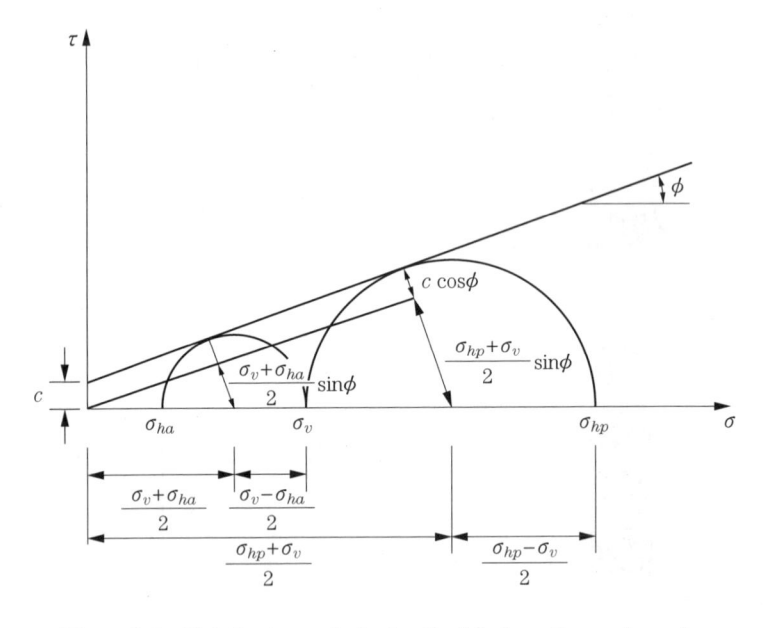

**Figure 9.6** Mohr's stress circle for Rankine's active and passive earth pressures

図 9.6 ランキンの主働土圧，受働土圧におけるモールの応力円

壁が後方に移動し背面が押される受働状態では，垂直土圧 $\sigma_v$ と水平土圧 $\sigma_h$ の関係が $\sigma_v < \sigma_h$ となると考え，伸張せん断破壊における水平土圧の大きさを受働土圧 $\sigma_{hp}$ とした（**図 9.5**）。

ランキンの土圧理論は，壁面は垂直で背面の地盤も水平と仮定し，地盤内の水平面および垂直面ではせん断応力が作用しない主応力面であると考えることで主働土圧，受働土圧を求めている。

垂直応力 $\sigma_v$ および主働，受働状態での水平応力 $\sigma_{ha}$, $\sigma_{hp}$ は主応力 $\sigma_1$, $\sigma_3$ であると考え，主応力におけるクーロンの破壊基準

$$\sigma_1(1-\sin\phi) = \sigma_3(1+\sin\phi) + 2c\cos\phi \tag{9.1}$$

を基に，主働土圧，受働土圧を求めてみる。

上述のように主働状態では，$\sigma_v > \sigma_{ha}$ であるので，$\sigma_v$ が最大主応力 $\sigma_1$, $\sigma_{ha}$ が最小主応力 $\sigma_3$ となる。一方，受働状態では，$\sigma_v < \sigma_{hp}$ であるので，$\sigma_{hp}$ が最大主応力 $\sigma_1$, $\sigma_v$ が最小主応力 $\sigma_3$ となる。このことを考慮し式(9.1)および**図 9.6** のモールの応力円より，主働土圧 $\sigma_{ha}$ および受働土圧 $\sigma_{hp}$ を求めると

$$\sigma_{ha} = \frac{1-\sin\phi}{1+\sin\phi}\sigma_v - 2c\frac{\cos\phi}{1+\sin\phi} \tag{9.2}$$

$$\sigma_{hp} = \frac{1+\sin\phi}{1-\sin\phi}\sigma_v + 2c\frac{\cos\phi}{1-\sin\phi} \tag{9.3}$$

となる。

ここで

$$\frac{1\mp\sin\phi}{1\pm\sin\phi} = \tan^2\left(\frac{\pi}{4}\mp\frac{\phi}{2}\right) \tag{9.4}$$

$$\frac{\mp\cos\phi}{1\pm\sin\phi} = \tan\left(\frac{\pi}{4}\mp\frac{\phi}{2}\right) \tag{9.5}$$

を考慮して，式(9.2), (9.3)を変形すると

$$\begin{pmatrix}\sigma_{ha}\\\sigma_{hp}\end{pmatrix} = \tan^2\left(\frac{\pi}{4}\mp\frac{\phi}{2}\right)\sigma_v \mp 2c\tan\left(\frac{\pi}{4}\mp\frac{\phi}{2}\right) \tag{9.6}$$

が求まる。

このとき，$K_a$, $K_p$ を

From Equations (9.2) and (9.3), we obtained,

$$\begin{pmatrix} \sigma_{ha} \\ \sigma_{hp} \end{pmatrix} = \tan^2\left(\frac{\pi}{4} \mp \frac{\phi}{2}\right) \sigma_v \mp 2c \tan\left(\frac{\pi}{4} \mp \frac{\phi}{2}\right) \tag{9.6}$$

Taking $K_a$ and $K_p$ as

$$K_a = \tan^2\left(\frac{\pi}{4} - \frac{\phi}{2}\right) \tag{9.7}$$

$$K_p = \tan^2\left(\frac{\pi}{4} + \frac{\phi}{2}\right) \tag{9.7'}$$

The relationship between the horizontal and vertical earth pressures are

$$\begin{pmatrix} \sigma_{ha} \\ \sigma_{hp} \end{pmatrix} = \begin{pmatrix} K_a \\ K_p \end{pmatrix} \sigma_v \mp 2c \begin{pmatrix} \sqrt{K_a} \\ \sqrt{K_p} \end{pmatrix} \tag{9.8}$$

Equation (9.8) is Rankine's equation for active and passive earth pressures. Rankine's coefficients $K_a$ and $K_p$ are called the "active earth pressure coefficient" and the "passive earth pressure coefficient", respectively.

For a sand without cohesion ($c=0$ kN/m²), the Equation (9.8) becomes

$$\begin{pmatrix} \sigma_{ha} \\ \sigma_{hp} \end{pmatrix} = \begin{pmatrix} K_a \\ K_p \end{pmatrix} \sigma_v \tag{9.9}$$

The resultant pressures for the active ($P_a$ [kN/m]) and passive ($P_p$ [kN/m]) earth pressures are

$$P_a = \frac{1}{2} \gamma_t H^2 K_a \tag{9.10}$$

$$P_p = \frac{1}{2} \gamma_t H^2 K_p \tag{9.11}$$

where, $\gamma_t$ is the unit weight of a soil and $H$ is the height of the ground measured from the toe of the retaining wall (Figure 9.5).

**【Example 9.1】** A 5 m high vertical retaining wall was constructed in a horizontal sandy soil as shown in **Figure 9.7**. Calculate the

$$K_a = \tan^2\left(\frac{\pi}{4} - \frac{\phi}{2}\right) \tag{9.7}$$

$$K_p = \tan^2\left(\frac{\pi}{4} + \frac{\phi}{2}\right) \tag{9.7'}$$

と置くと，式(9.6)の垂直土圧と水平土圧との関係は

$$\begin{pmatrix} \sigma_{ha} \\ \sigma_{hp} \end{pmatrix} = \begin{pmatrix} K_a \\ K_p \end{pmatrix} \sigma_v \mp 2c \begin{pmatrix} \sqrt{K_a} \\ \sqrt{K_p} \end{pmatrix} \tag{9.8}$$

で表される。

 式(9.8)は，ランキンによる主働土圧，受働土圧を求める式で，$K_a$，$K_p$ は，それぞれランキンの主働土圧係数および受働土圧係数と呼ばれる係数である。ところで，粘着力のない（$c = 0\ \mathrm{kN/m^2}$）砂の場合，式(9.8)は

$$\begin{pmatrix} \sigma_{ha} \\ \sigma_{hp} \end{pmatrix} = \begin{pmatrix} K_a \\ K_p \end{pmatrix} \sigma_v \tag{9.9}$$

で表され，このときの主働土圧合力 $P_a$〔kN/m〕，受働土圧合力 $P_p$〔kN/m〕は

$$P_a = \frac{1}{2}\gamma_t H^2 K_a \tag{9.10}$$

$$P_p = \frac{1}{2}\gamma_t H^2 K_p \tag{9.11}$$

となる。

 ここで，$\gamma_t$：土の単位体積重量，$H$：地盤の高さである。

【例題 9.1】 図9.7に示す水平な砂地盤に設置された，高さ5 m の垂直な擁壁に作用する主働土圧合力 $P_a$ および受働土圧合力 $P_p$ を，ランキンの土圧

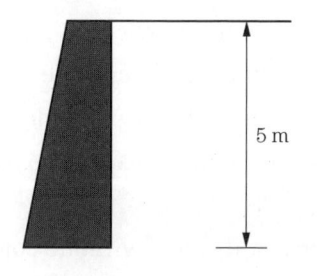

**Figure 9.7** A retaining wall constructed in a horizontal sandy soil
図9.7 水平な砂地盤に設置された擁壁

active $P_a$ and passive $P_p$ earth pressures on the wall by using Rankine's earth pressure theory. Take the unit weight of sand as ($\gamma_t = 18$ kN/m³) and the angle of internal friction as ($\phi = 30°$).

**Solution**  From Equations (9.10) and (9.11), Rankine's active $P_a$ and passive $P_p$ earth pressures for a sandy soil without cohesion are as follows:

$$P_a = \frac{1}{2}\gamma_t H^2 K_a \text{ and } P_p = \frac{1}{2}\gamma_t H^2 K_p$$

Now, calculating the active and passive earth pressure coefficients for ($\phi = 30°$), $K_a$ and $K_p$ are.

$$K_a = \tan^2\left(\frac{\pi}{4} - \frac{\phi}{2}\right) = \tan^2 30 = 0.333$$

$$K_p = \tan^2\left(\frac{\pi}{4} + \frac{\phi}{2}\right) = \tan^2 60 = 3$$

Therefore, $P_a$ and $P_p$ are given by

$$P_a = \frac{1}{2}\gamma_t H^2 K_a = \frac{1}{2} \times 18 \times 5^2 \times 0.333 = 75 \text{ kN/m}$$

$$P_p = \frac{1}{2}\gamma_t H^2 K_p = \frac{1}{2} \times 18 \times 5^2 \times 3 = 675 \text{ kN/m}$$

## 9.3  Coulomb's Earth Pressure

Rankine's earth pressure theory assumed that only principal stresses acted on the soil behind a vertical wall in a horizontal ground. Equations for active and passive earth pressures were derived for stress conditions that satisfied all the failure conditions of a soil mass in the ground. However, in practice, the surfaces of the wall and soil surrounding the wall are inclined (**Figure 9.8**), not vertical.

Therefore, an appropriate theory must also consider the geometry of the wall and its affect on the slope of the soil. Coulomb's earth pressure theory examines the effect of inclined walls on earth pressure.

Coulomb (1776) considered a sandy soil mass without cohe-

理論から求めよ。なお，砂の単位体積重量は $\gamma_t=18\,\mathrm{kN/m^3}$，内部摩擦角は $\phi=30°$ とする。

**解答** 粘着力のない砂地盤におけるランキンの主働土圧合力 $P_a$，受働土圧合力 $P_p$ は，式(9.10)，(9.11)より $P_a=\dfrac{1}{2}\gamma_t H^2 K_a$，$P_p=\dfrac{1}{2}\gamma_t H^2 K_p$ で表される。

$\phi=30°$ における主働土圧係数 $K_a$，受働土圧係数 $K_p$ を求めると

$$K_a=\tan^2\left(\dfrac{\pi}{4}-\dfrac{\phi}{2}\right)=\tan^2 30=0.333$$

$$K_p=\tan^2\left(\dfrac{\pi}{4}+\dfrac{\phi}{2}\right)=\tan^2 60=3$$

であるので

$$P_a=\dfrac{1}{2}\gamma_t H^2 K_a=\dfrac{1}{2}\times 18\times 5^2\times 0.333=75\,\mathrm{kN/m}$$

$$P_p=\dfrac{1}{2}\gamma_t H^2 K_p=\dfrac{1}{2}\times 18\times 5^2\times 3=675\,\mathrm{kN/m}$$

となる。

## 9.3 クーロン土圧

ランキンの土圧理論では，壁が鉛直で背面の地盤が水平な条件を仮定するこ

---

**Christian Otto Mohr**（オットー・モール，1835〜1918）

ドイツの土木工学者。ハノーヴァーにある Polytechnic school で学ぶ。

まず，ハノーヴァーとオルデンブルグで鉄道技師として勤務し，橋や鋼製トラスの設計に携わった。しばらくしてシュトゥットゥガルト・ポリテクニークの教授，ドレスデン・ポリテクニークの教授になった。

1882年，モールの応力円として有名な図解法を開発し，それをせん断応力に基づく初期の強度理論に用いた。

**図14** オットー・モール

sion that is bounded by a wall with a slip surface as shown in Figure 9.8. He assumed that the soil mass moves along the slip surface caused by the movement of the wall. Earth pressure on the wall surface were calculated using an equilibrium approach where the force from the weight $W$ of soil mass (ABC) and the reaction forces $P_a$ and $R$ on surfaces (AB) and (BC) are balanced. When the wall moves forward, the active earth pressure $P_a$ acts on the surface of the wall. The balance of forces is shown by the force triangle given in Figure 9.8. Applying the sine rule,

$$\frac{P_a}{\sin(\alpha-\phi)} = \frac{W}{\sin(\theta+\delta-\alpha+\phi)} \qquad (9.12)$$

and therefore, $P_a$ is given by

$$P_a = \frac{\sin(\alpha-\phi)}{\sin(\theta+\delta-\alpha+\phi)} W \qquad (9.13)$$

Differentiating $P_a$ with respect to $\alpha$, maximum value of $W$ is

$$W = \frac{1}{2} \overline{BC}^2 \gamma_t \frac{\sin(\alpha-\phi)\sin(\theta+\delta-\alpha+\phi)}{\sin(\theta+\delta)} \qquad (9.14)$$

where, $\gamma_t$ is the unit weight of soil.

The area of the triangle $\triangle$BCE in **Figure 9.9** is given by

$$\text{Area of } \triangle BCE = \frac{1}{2} \overline{BC}^2 \frac{\sin(\alpha-\phi)\sin(\theta+\delta-\alpha+\phi)}{\sin(\theta+\delta)} \qquad (9.15)$$

From Equation (9.15), the $W$ of Equation (9.14) is calculated as $W =$ Area of A $\triangle$BCE $\times \gamma_t$.

$W$ becomes

$$W = \frac{1}{2} \overline{BE} \times \overline{CF} \times \gamma_t \qquad (9.16)$$

Triangle BCE in Figure 9.9 and the force triangle $P_a$, $W$, $R$ of Figure 9.8 are similar. Thus,

$$\frac{\overline{BE}}{W} = \frac{\overline{CE}}{P_a} \qquad (9.17)$$

and $P_a$ becomes

## 9.3 Coulomb's Earth Pressure

とで主応力と考え，地盤内のすべてで土の破壊条件を満足する応力状態を想定して主働土圧，受働土圧を求める式を導いた。しかし，実際には壁面や背面地盤が傾斜している場合もあり，このような条件に対して適用できる理論が必要となる。

クーロンは，**図 9**.8(a)に示すような壁面とすべり面で囲まれた粘着力のない砂の土塊が，壁の移動によってすべり面に沿って移動すると仮定し，土塊 ABC の重量 $W$ と，AB 面および BC 面に作用する反力 $P_a$, $R$ とが釣合いを保つと考えて，壁面に作用する土圧を求めた。

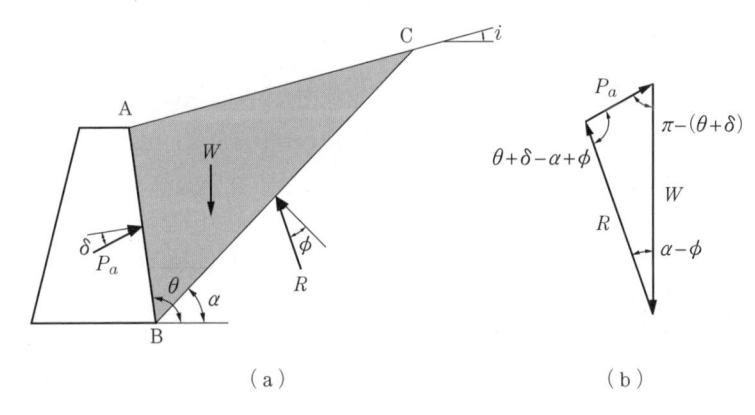

**Figure 9**.8 Assumptions for Coulomb's earth pressure and the force triangle
**図 9**.8 クーロンの土圧の仮定と力の三角形

壁が前方に移動する場合の壁面に及ぼす主働土圧合力 $P_a$ を，図(b)に示す力の釣合い三角形を基に正弦定理を適用して求めると

$$\frac{P_a}{\sin(\alpha-\phi)} = \frac{W}{\sin(\theta+\delta-\alpha+\phi)} \tag{9.12}$$

となり，$P_a$ は

$$P_a = \frac{\sin(\alpha-\phi)}{\sin(\theta+\delta-\alpha+\phi)} W \tag{9.13}$$

で表される。

このとき，$\alpha$ について微分して極値を求め，$P_a$ を最大にするときの $W$ を求

## Chapter 9　Earth Pressure

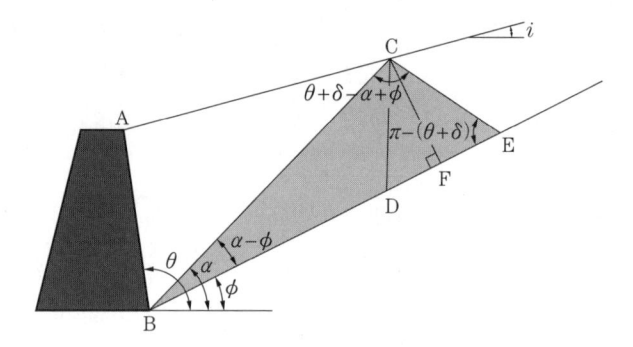

**Figure 9.9**　Calculation of Coulomb's earth pressure
図 9.9　クーロンの土圧の計算

$$P_a = \frac{\overline{CE}}{\overline{BE}} W \tag{9.18}$$

Substituting Equation (9.16) into Equation (9.18) yields

$$P_a = \frac{1}{2} \overline{CE} \times \overline{CF} \times \gamma_t \tag{9.19}$$

On BE, taking $\overline{DE}$ as an equivalent length of $\overline{CE}$, gives

$$P_a = \frac{1}{2} \overline{DE} \times \overline{CF} \times \gamma_t \tag{9.20}$$

$$= \text{Area of } \triangle CDE \times \gamma_t \tag{9.21}$$

Here, the Area of $CDE = \frac{1}{2} \overline{DE} \times \overline{CF} = \frac{1}{2} \overline{CE}^2 \sin(\theta + \delta)$

$$\tag{9.22}$$

Then $\overline{CE}$ is given as follows

$$\overline{CE} = \frac{1}{1 + \sqrt{\dfrac{\sin(\theta + \delta) \sin(\phi - i)}{\sin(\theta + \delta) \sin(\theta - i)}}} \frac{\sin(\theta - \phi)}{\sin(\theta + \delta)} \frac{H}{\sin \theta} \tag{9.23}$$

Calculating the resultant active earth pressure $P_a$ from Equations (9.22) and (9.21):

$$P_a = \frac{1}{2} \gamma_t H^2 K_a \tag{9.24}$$

where $K_a$ is the coefficient of active earth pressure and is given by

めると

$$W = \frac{1}{2}\overline{BC}^2 \gamma_t \frac{\sin(\alpha-\phi)\sin(\theta+\delta-\alpha+\phi)}{\sin(\theta+\delta)} \tag{9.14}$$

となる。

ここで，$\gamma_t$ は土の単位体積重量である。

図 9.9 において，△BCE の面積は

$$\triangle \text{BCE の面積} = \frac{1}{2}\overline{BC}^2 \frac{\sin(\alpha-\phi)\sin(\theta+\delta-\alpha+\phi)}{\sin(\theta+\delta)} \tag{9.15}$$

である。

式(9.15)より，式(9.14)の $W$ は，$W = \triangle$BCE の面積 $\times \gamma_t$ となり

$$W = \frac{1}{2}\overline{BE} \times \overline{CF} \times \gamma_t \tag{9.16}$$

で表される。

△BCE と図 9.8 に示す $P_a$, $W$, $R$ の合力の三角形は相似であるので

$$\frac{\overline{BE}}{W} = \frac{\overline{CF}}{P_a} \tag{9.17}$$

となり

$$P_a = \frac{\overline{CE}}{\overline{BE}} W \tag{9.18}$$

である。

式(9.18)に式(9.16)を代入すると

$$P_a = \frac{1}{2}\overline{CE} \times \overline{CF} \times \gamma_t \tag{9.19}$$

となる。

$\overline{CE}$ と長さが等しい $\overline{DE}$ を BE 上にとると

$$P_a = \frac{1}{2}\overline{DE} \times \overline{CF} \times \gamma_t \tag{9.20}$$

$$= \triangle \text{CDE の面積} \times \gamma_t \tag{9.21}$$

である。

ここで

$$K_a = \left[ \frac{\sin(\theta-\phi)}{\sin\theta \left\{ \sqrt{\sin(\theta+\delta)} + \sqrt{\frac{\sin(\theta+\delta)\sin(\phi-i)}{\sin(\theta-i)}} \right\}} \right]^2 \quad (9.25)$$

For the passive earth pressure case (i.e., when a wall moves backward), the resultant passive earth pressure $P_p$ on the wall surface is

$$P_p = \frac{1}{2} \gamma_t H^2 K_p \quad (9.26)$$

where, $K_p$ is the coefficient of passive earth pressure that is given by

$$K_p = \left[ \frac{\sin(\theta+\phi)}{\sin\theta \left\{ \sqrt{\sin(\theta-\delta)} - \sqrt{\frac{\sin(\theta+\delta)\sin(\phi+i)}{\sin(\theta-i)}} \right\}} \right]^2 \quad (9.27)$$

$K_a$ and $K_p$ of Equations (9.25) and (9.27) are known as Coulomb's earth pressure coefficients for active and passive conditions, respectively. These coefficients can be determined from the angle of internal friction $\phi$, the frictional angle of the wall's surface $\delta$, the wall's inclination angle $\theta$ and the inclination of the ground, angle $i$.

【Example 9.2】 A 5 m high vertical retaining wall was constructed in a sandy soil with ground angle (slope) of 10° as shown in **Figure 9.10**. Calculate the resultant active and passive earth pressures $P_a$ and $P_p$ by using Coulomb's earth pressure theory. Take the unit weight of sand as ($\gamma_t = 18$ kN/m³), the angle of internal friction as ($\phi = 30°$) and the frictional angle of wall's surface as ($\delta = 20°$).

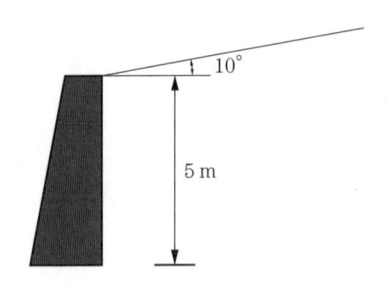

**Figure 9.10** A vertical retaining wall constructed in sandy soil with 10° ground angle (slope)
図 9.10 10°傾斜した砂地盤に設置された擁壁

$$\triangle \text{CDE の面積} = \frac{1}{2}\overline{\text{DE}} \times \overline{\text{CF}} = \frac{1}{2}\overline{\text{CE}}^2 \sin(\theta + \delta) \tag{9.22}$$

であり，$\overline{\text{CE}}$ は

$$\overline{\text{CE}} = \frac{1}{1+\sqrt{\dfrac{\sin(\theta+\delta)\sin(\phi-i)}{\sin(\theta+\delta)\sin(\theta-i)}}} \frac{\sin(\theta-\phi)}{\sin(\theta+\delta)} \frac{H}{\sin\theta} \tag{9.23}$$

で示されるので，式(9.21)，(9.22)から主働土圧合力 $P_a$ を求めると

$$P_a = \frac{1}{2}\gamma_t H^2 K_a \tag{9.24}$$

となる．ここで，$K_a$ は主働土圧係数で

$$K_a = \left[\frac{\sin(\theta-\phi)}{\sin\theta\left\{\sqrt{\sin(\theta+\delta)} + \sqrt{\dfrac{\sin(\theta+\delta)\sin(\phi-i)}{\sin(\theta-i)}}\right\}}\right]^2 \tag{9.25}$$

で示される．

　壁が後方に移動するとき壁面に及ぼす受働土圧合力 $P_a$ を，主働土圧合力の場合と同様にして求めと

$$P_p = \frac{1}{2}\gamma_t H^2 K_p \tag{9.26}$$

となる．ここで，$K_p$ は受働土圧係数で

$$K_p = \left[\frac{\sin(\theta+\phi)}{\sin\theta\left\{\sqrt{\sin(\theta-\delta)} - \sqrt{\dfrac{\sin(\theta+\delta)\sin(\phi+i)}{\sin(\theta-i)}}\right\}}\right]^2 \tag{9.27}$$

で示される．

　式(9.25)，(9.27)の $K_a$，$K_p$ はクーロンの主働土圧係数および受働土圧係数と呼ばれる係数で，地盤の内部摩擦角 $\phi$，壁面の粗さ角 $\delta$，壁の傾斜角度 $\theta$，地盤の傾斜角 $i$ によって求めることができる．

【例題 9.2】 図9.10 に示す，10°傾斜した砂地盤に設置された，高さ5mの垂直な擁壁に作用する主働土圧合力 $P_a$，受働土圧合力 $P_p$ を，クーロンの土圧理論から求めよ．なお，砂の単位体積重量は $\gamma_t = 18\,\text{kN/m}^3$，内部摩擦角

## Chapter 9 Earth Pressure

**Solution**  From Equations (9.23) and (9.25), the resultant Coulomb's active and passive earth pressures ($P_a$ and $P_p$) can be calculated from

$$P_a = \frac{1}{2}\gamma_t H^2 K_a, \quad P_p = \frac{1}{2}\gamma_t H^2 K_p$$

where, $K_a$ is the active earth pressure coefficient and $K_p$ is the passive earth pressure coefficient. $K_a$ and $K_p$ are calculated by using Equations (9.24) and (9.26) as follows:

$$K_a = \left[ \frac{\sin(\theta-\phi)}{\sin\theta \left\{ \sqrt{\sin(\theta+\delta)} + \sqrt{\frac{\sin(\theta+\delta)\sin(\phi-i)}{\sin(\theta-i)}} \right\}} \right]^2$$

$$= \left[ \frac{\sin(90-30)}{\sin 90 \times \left\{ \sqrt{\sin(90+20)} + \sqrt{\frac{\sin(30+20)\sin(30-10)}{\sin(90-10)}} \right\}} \right]^2$$

$$= 0.34$$

$$K_p = \left[ \frac{\sin(\theta+\phi)}{\sin\theta \left\{ \sqrt{\sin(\theta-\delta)} - \sqrt{\frac{\sin(\phi+\delta)\sin(\phi+i)}{\sin(\theta-i)}} \right\}} \right]^2$$

$$= \left[ \frac{\sin(90+30)}{\sin 90 \times \left\{ \sqrt{\sin(90-20)} - \sqrt{\frac{\sin(30+20)\sin(30+10)}{\sin(90-10)}} \right\}} \right]^2$$

$$= 10.93$$

Therefore, the resultant Coulomb's active and passive earth pressures ($P_a$ and $P_p$) are

$$P_a = \frac{1}{2}\gamma_t H^2 K_a = \frac{1}{2} \times 18 \times 5^2 \times 0.34 = 76.5 \text{ kN/m}$$

$$P_p = \frac{1}{2}\gamma_t H^2 K_p = \frac{1}{2} \times 18 \times 5^2 \times 10.93 = 2\,459.25 \text{ kN/m}$$

は $\phi=30°$,擁壁面の粗さ角は,$\delta=20°$ とする。

**解答** クーロンの主働土圧合力 $P_a$,受働土圧合力 $P_p$ は,式(9.23),(9.25),

$$P_a = \frac{1}{2}\gamma_t H^2 K_a, \quad P_p = \frac{1}{2}\gamma_t H^2 K_p$$

により求めることができる。

ここで,主働土圧係数 $K_a$,受働土圧係数 $K_p$ を式(9.24),(9.26)から求めると

$$K_a = \left[\frac{\sin(\theta-\phi)}{\sin\theta\left\{\sqrt{\sin(\theta+\delta)} + \sqrt{\dfrac{\sin(\theta+\delta)\sin(\phi-i)}{\sin(\theta-i)}}\right\}}\right]^2$$

$$= \left[\frac{\sin(90-30)}{\sin 90 \times \left\{\sqrt{\sin(90+20)} + \sqrt{\dfrac{\sin(30+20)\sin(30-10)}{\sin(90-10)}}\right\}}\right]^2$$

$$= 0.34$$

$$K_p = \left[\frac{\sin(\theta+\phi)}{\sin\theta\left\{\sqrt{\sin(\theta-\delta)} - \sqrt{\dfrac{\sin(\phi+\delta)\sin(\phi+i)}{\sin(\theta-i)}}\right\}}\right]^2$$

$$= \left[\frac{\sin(90+30)}{\sin 90 \times \left\{\sqrt{\sin(90-20)} - \sqrt{\dfrac{\sin(30+20)\sin(30+10)}{\sin(90-10)}}\right\}}\right]^2$$

$$= 10.93$$

となる。

したがって,クーロンの主働土圧合力 $P_a$,受働土圧合力 $P_p$ は

$$P_a = \frac{1}{2}\gamma_t H^2 K_a = \frac{1}{2}\times 18 \times 5^2 \times 0.34 = 76.5 \text{ kN/m}$$

$$P_p = \frac{1}{2}\gamma_t H^2 K_p = \frac{1}{2}\times 18 \times 5^2 \times 10.93 = 2\,459.25 \text{ kN/m}$$

である。

# Chapter 10  Bearing Capacity

## 10.1  Bearing Capacity Formula

When a structure is constructed on a ground, the load from the structure creates stresses in the ground. If the stress that developed in the ground exceeds the strength of the soil, then the ground deforms or fails. Thus, it is necessary to determine the bearing capacity of a ground when designing foundations such as footings and pilings.

Load-settlement relationships vary depending on the ground conditions when a load is applied to a horizontal ground as shown in **Figure 10.1**. The ultimate bearing capacity of a dense-sandy ground or a firm-clayey ground can be determined precisely when the settlement increases suddenly. In this case, development of a slip surface (shear surface) is often observed because of plastic failure of the ground. Failure patterns for this type of failure are either an "general shear failure" or a "local shear failure". For a "general shear failure", the slip surface extends underneath and pass the foundation as shown in **Figure 10.2** (a). In contrast, for a "local shear failure", the plastic failure zone is localized to zones beneath the foundation as shown in Figure 10.2 (b). Local shear failure often occurs for a loose-sandy ground or a soft-clayey ground, and the ultimate bearing capacity is difficult to determine, precisely, because the settlement occurs slowly during and after loading.

Now, let's examine the load-settlement curves for an general shear failure (Figure 10.1). Initially, elastic behavior is observed

# 10章 支 持 力

## 10.1 支 持 力 公 式

　地盤上に構造物などを建設する場合，構造物の荷重によって発生する地盤内の応力が地盤のもっている強度を超えると，地盤は破壊あるいは変形する。このため，フーチングや杭などの基礎構造物を設計する場合，地盤がどの程度の支持力をもつか求めておくことが重要である。

　水平な地盤に載荷を行った場合，地盤状態によって図10.1のように荷重-沈

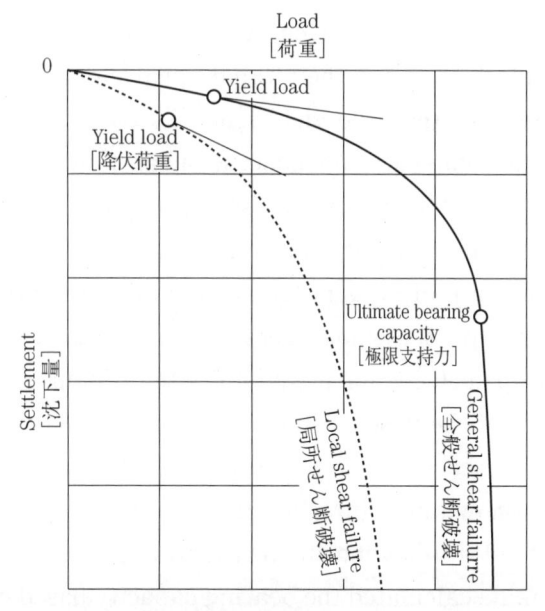

**Figure 10.1** Load-settlement curves and ultimate value
図10.1 極限荷重-沈下曲線

(a) General shear failure

(b) Local shear failure

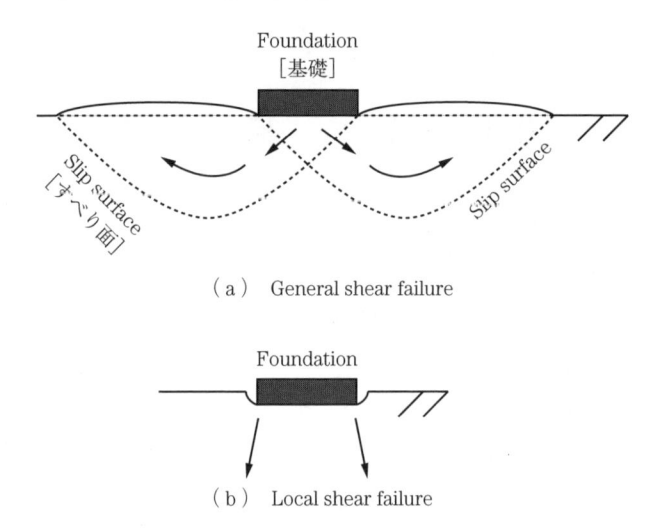

**Figure 10.2** General shear failure and local shear failure
図 10.2 全般せん断破壊と局所せん断破壊

for a very small range of loads as indicated by the straight line portion of the load-settlement curves. After this elastic behavior, a soil behaves plastically and the ultimate bearing capacity occurs for loads much greater than the yield load. To determine a theoretical ultimate bearing capacity, an general shear failure is assumed to result in the development of slip surface in the ground. For this purpose, the bearing capacity formula is used.

In the case of an general shear failure, the failure patterns differ depending on the smoothness or roughness of the base of a foundation. Slip surfaces for a smooth base develop on both the left and right sides of the base as shown in **Figure 10.3** (a). On the other hand, for a rough base, a wedge is formed just beneath the base of the foundation as shown in Figure 10.3 (b). Slip surfaces develop on both left and right sides following the penetration of the wedge.

Rankine calculated the bearing capacity based on the failure patterns for a smooth base for a horizontal ground as shown in

下関係が異なる。密な砂地盤や硬い粘土地盤の場合，沈下量が急激に増大し極限支持力を明確に求めることができる。この場合，地盤は塑性破壊を起こし，すべり面（せん断面）の発達がみられ，地盤内の破壊パターンは，**図 10.2 (a)** のような地盤全体にすべり面が発達する全般せん断破壊と考えることができる。

これに対し，緩い砂地盤や柔らかい粘土地盤の場合，載荷直下が圧縮され沈下がだらだらと進み，極限支持力は明確に求まらない。この場合，地盤内の破壊パターンは，図 10.2(b) のように，塑性破壊が地盤全体に広がらない局所せん断破壊と考えることができる。

図 10.1 の荷重-沈下曲線において全般せん断破壊の荷重変化をみると，ごく初期に荷重と沈下が線形関係を示す弾性的な挙動を示し，その後塑性的な挙動を示し，降伏荷重を超えて最終的に極限支持力を示す。極限支持力を理論的に求める場合，地盤内におけるすべり面の発達状況を仮定できる全般せん断破壊

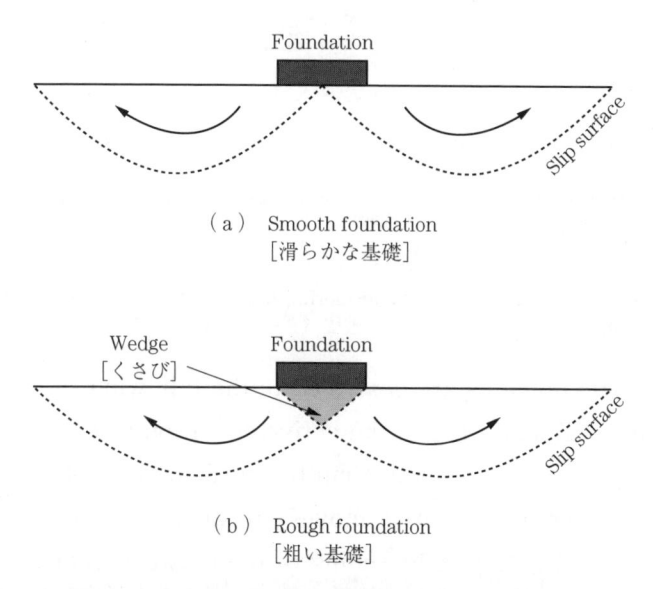

（a） Smooth foundation
［滑らかな基礎］

（b） Rough foundation
［粗い基礎］

**Figure 10.3** Differences in the failure patterns for smooth and rough bases
図 10.3　基礎底面の滑・粗による破壊パターンの違い

**Figure 10.4.** The ground is loaded by a footing of width $B$ and a load $Q$ along with uniform loads on the ground surface (Figure 10.4). Slip surfaces develop symmetrically about the center of the footing. The shape of the slip surface is formed by two straight lines (OC, CD on the right and OC′, C′D′ on the left). Assuming that line (AC) is the boundary of the plastic domain on the right side of the footing, the region from the center of the footing and to the right of the footing can be divided into two domains, i.e., (AOC) (Domain I) and (ACD) (Domain II).

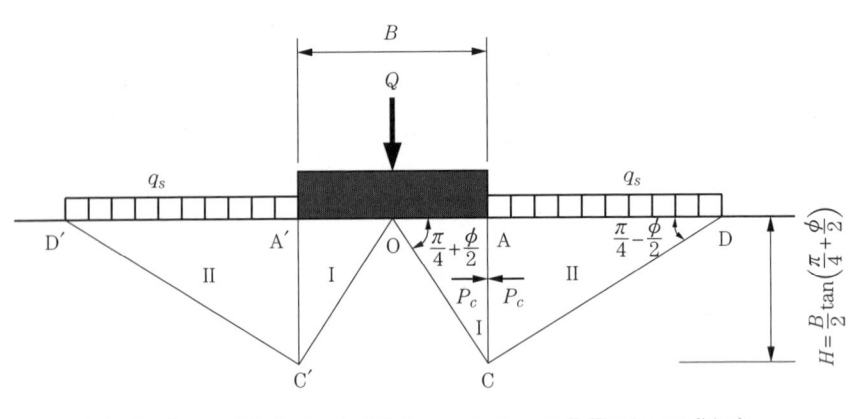

Domain II (Passive condition)   Domain I (Active condi-   Domain II (Passive condition)
領域 II (受働状態)                tion)                     領域 II (受働状態)
                                領域 I (主働状態)

**Figure 10.4**  Bearing capacity considering failure pattern of a smooth base
図 10.4  底面が滑らかな場合の破壊パターンを基にした支持力

Both plastic domains are described in the following.

Domain I : Domain I is in the active condition where the vertical stress is greater than the horizontal stress because the boundary (AC) moves towards the right side that results in a loosening of the plastic domain. Here, the failure surface forms an angle of $(\pi/4 + \phi/2)$ with the ground surface.

を対象とする必要があり，この場合の支持力公式を考えてみる．

全般せん断破壊を考える場合，基礎の底面が滑らかな場合と粗い場合で破壊パターンが異なることが示されており，底面が滑らかな場合には，図10.3(a)のように基礎中央から左右にすべり面が発達するのに対し，底面が粗い場合には，図(b)のように基礎直下にくさびを形成し，このくさびの貫入に伴って左右のすべり面が発達すると考えられる．

ランキンは，地盤が水平で底面が滑らかな場合の破壊パターンを基に支持力を求めた．図10.4に示す地盤に幅 $B$ の基礎が荷重 $Q$ で載荷するとともに，地盤表面に $q_s$ の等分布荷重が載荷するとき，基礎中央から左右対称のすべり面が発達し，このときのすべり面の形状は，基礎中央から左右対象な2本の直線（OC，CD および OC′，C′D′）からなると仮定した．このとき，基礎中央から右側の領域を考えると，AC を壁と仮定し，AOC（領域 I）および ACD（領域 II）の塑性領域に分けられると考えた．

両塑性領域は，

領域 I：この領域は，AC が右側に移動し塑性領域は緩むため，垂直応力が水平応力より大きい主働状態となり，地盤面と破壊面とは $\pi/4 + \phi/2$ の角度をなす．

領域 II：この領域は，AC が右側に移動し塑性領域は押されるため，垂直応力が水平応力より小さい受働状態となり，地盤面と破壊面とは $\pi/4 - \phi/2$ の角度をなす．

と仮定した．

ここで，AC 面に作用する合力を $P_c$ とすると，

領域 I では

$$\text{垂直応力}：\sigma_v = \frac{Q}{B} + \frac{\gamma_t H}{2}, \quad \text{水平応力}：\sigma_h = \frac{P_c}{H} \tag{10.1}$$

領域 II では

$$\text{垂直応力}：\sigma_v = q_s + \frac{\gamma_t H}{2}, \quad \text{水平応力}：\sigma_h = \frac{P_c}{H} \tag{10.2}$$

**Domain II** : Domain II is in the passive condition where the vertical stress is less than the horizontal stress because the boundary (AC) moves towards right side and compresses the plastic domain. Here, the failure surface forms an angle of $(\pi/4-\phi/2)$ with the ground surface. Now consider that the resultant $P_c$ is acting on surface (AC):

In Domain I, the stresses are as follows:

The vertical stress is $\sigma_v = \dfrac{Q}{B} + \dfrac{\gamma_t H}{2}$ and the horizontal stress is

$$\sigma_h = \dfrac{P_c}{H} \tag{10.1}$$

In Domain II, the stresses are as follows:

The vertical stress is $\sigma_v = q_s + \dfrac{\gamma_t H}{2}$ and for the horizontal stress is

$$\sigma_h = \dfrac{P_c}{H} \tag{10.2}$$

From Rankine's earth pressure theory, the active earth pressure coefficient $K_a$ and the passive earth pressure coefficient $K_p$ are given by

$$\begin{pmatrix} \sigma_{ha} \\ \sigma_{hp} \end{pmatrix} = \begin{pmatrix} K_a \\ K_p \end{pmatrix} \sigma_v \mp 2c \begin{pmatrix} \sqrt{K_a} \\ \sqrt{K_p} \end{pmatrix} \tag{10.3}$$

Substituting Equations (10.1) and (10.2) into Equation (10.3), yields: for Domain I in the active condition:

$$\dfrac{P_c}{H} = K_a \left( \dfrac{Q}{B} + \dfrac{\gamma_t H}{2} \right) - 2c\sqrt{K_a} \tag{10.4}$$

for Domain II in the passive condition:

$$\dfrac{P_c}{H} = K_p \left( q_s + \dfrac{\gamma_t H}{2} \right) + 2c\sqrt{K_p} \tag{10.5}$$

Considering $K_a \times K_p = 1$ and eliminating $P_c$ from Equations (10.4) and (10.5), the following equation is obtained:

となる。

主働土圧係数を $K_a$,受働土圧係数を $K_p$ とすると,ランキン土圧より

$$\begin{pmatrix}\sigma_{ha}\\ \sigma_{hp}\end{pmatrix}=\begin{pmatrix}K_a\\ K_p\end{pmatrix}\sigma_v\mp 2c\begin{pmatrix}\sqrt{K_a}\\ \sqrt{K_p}\end{pmatrix} \qquad (10.3)$$

であり,式(10.3)に式(10.1),(10.2)を代入すると,

主働状態の領域 I は

$$\frac{P_c}{H}=K_a\left(\frac{Q}{B}+\frac{\gamma_t H}{2}\right)-2c\sqrt{K_a} \qquad (10.4)$$

受働状態の領域 II は

$$\frac{P_c}{H}=K_p\left(q_s+\frac{\gamma_t H}{2}\right)+2c\sqrt{K_p} \qquad (10.5)$$

---

### Karl von Terzaghi（テルツァーギ,1883〜1963）

　チェコのプラハ生まれ。土質力学の父といわれる。筆者の一人が学生時代学んだ赤井浩一先生の「土質力学」の序には「土質力学が近代的な学問として世に認められるようになったのは,テルツァーギの著書 Erdbaumechanik の発刊（1925）を契機とするのは衆目の一致するところである」とある。

　父親の退職後,家族はオーストリアのグラーツに移った。1900年,今のグラーツ工科大学に入学,機械工学を学んだが,学生時代は手に負えない者だったようだ。1年間軍隊に入れさせられ,そのとき英語の地質のマニュアルをドイツ語に訳したのが後の彼を作ったようだ。

図15　テルツァーギ

図16　テルツァーギが作った最初の圧密試験器。ノルウェー地質学研究所（NGI）のテルツァーギ図書室にて（1995年,勝山撮影）

$$\frac{Q}{B} = \frac{\gamma_t}{2} BN_r + cN_c + q_s N_q \qquad (10.6)$$

Equation (10.6) is called Rankine's bearing capacity formula. Each term in this formula is defined as follows:

$\frac{\gamma_t}{2} BN_r$ : Frictional resistance exhibited by the soil mass

$cN_c$ : Resisting force of cohesion along slip surface

$q_s N_q$ : Frictional resistance developed by surcharge $q_s$

where, $N_r$, $N_c$ and $N_q$ are the coefficients of bearing capacity that are given by

$$N_r = \frac{1}{2}(K_p^{2.5} - K_p^{0.5}) \qquad (10.7)$$

$$N_c = 2(K_p^{1.5} + K_p^{0.5}) \qquad (10.8)$$

$$N_q = K_p^2 \qquad (10.9)$$

The bearing capacity coefficients $N_r$, $N_c$ and $N_q$ can be calculated from Rankine's active earth pressure coefficient $K_p = \tan^2\left(\frac{\pi}{4} + \frac{\phi}{2}\right)$, and thus, the bearing capacity coefficients are functions of the angle of internal friction $\phi$.

## 10.2 Shallow Foundation

A foundation transfers the loads from a structure to the ground and can be thought of as two parts, namely, a shallow foundation and a deep foundation. The division into two parts is based on the ratio of the width $B$ and depth $D_f$ of the foundation. A shallow foundation is one that is constructed to a relatively shallow depth with $D_f/B$ being less than 1 (one). Formulations of bearing capacity equations for a shallow foundation have been proposed by different researchers such as Terzaghi, Prandtl and Meyerhof, etc. The formulations are basically the same for each approach except for the assumptions related

となる。

$K_a \times K_p = 1$ となることを考慮し，式(10.4)，式(10.5)より $P_c$ を消去すると

$$\frac{Q}{B} = \frac{\gamma_t}{2} BN_r + cN_c + q_s N_q \tag{10.6}$$

が求まる。

式(10.6)はランキンの支持力公式と呼ばれ，各項目は，

$\frac{\gamma_t}{2} BN_r$ の項目：土の自重によって発揮される摩擦抵抗力

$cN_c$ ：すべり面に沿って働く粘着力による抵抗力

$q_s N_q$ ：サーチャージ $q_s$ の押えによって生じる摩擦抵抗

である。

ここで，$N_r$, $N_c$, $N_q$ は支持力係数と呼ばれる係数で，それぞれ

$$N_r = \frac{1}{2}(K_p^{2.5} - K_p^{0.5}) \tag{10.7}$$

$$N_c = 2(K_p^{1.5} + K_p^{0.5}) \tag{10.8}$$

$$N_q = K_p^2 \tag{10.9}$$

で求められる。

ところで，支持力係数 $N_r$, $N_c$, $N_q$ は，ランキンの受働土圧係数

$$K_p = \tan^2\left(\frac{\pi}{4} + \frac{\phi}{2}\right)$$

から求めることができ，内部摩擦角 $\phi$ の関数となる。

## 10.2 浅 い 基 礎

構造物の荷重を地盤に伝える基礎は，基礎幅 $B$ と基礎の根入れ深さ $D_f$ の比によって，浅い基礎，深い基礎に分けられる。このうち，浅い基礎は $D_f/B$ が1以下の比較的浅い位置に施工される基礎である。浅い基礎に対する支持力公式は，テルツァーギ，プラントル，マイヤーホフなど多くのものが提案されており，各公式は破壊面の形状などの仮定が異なるものの，基本的な考え方に

to the shape of the failure surface.

Terzaghi's bearing capacity formula is used most often and assumes an general shear failure as shown in **Figure 10.5**. The plastic domain is divided into an active wedge domain I, a transition domain II and a passive domain III. The ultimate bearing capacity is calculated by assuming that the foundation with Domain I penetrated together. Therefore, a passive domain III is formed on the boundary of domain II.

The ultimate bearing capacity based on Terzaghi's approach is given by Equation (10.10) for a ground loaded by a footing of width $B$ and a load $Q$ along with uniform loads on the ground surface.

$$\frac{Q}{B}(=q)=\frac{\gamma_t}{2}BN_r+cN_c+q_s N_q \tag{10.10}$$

$$=\frac{\gamma_t}{2}BN_r+cN_c+\gamma_{t2} D_f N_q \tag{10.10'}$$

where $q$ is the intensity of ultimate bearing capacity of ground [kN/m²], $\gamma_t$ is the unit weight of ground [kN/m³], $B$ is the width of foundation [m], $c$ is the cohesion [kN/m²], $q_s$ is the uniform loads on ground surface (surcharge loads) [kN/m²], and $N_r$, $N_c$ and $N_q$ are the coefficients of bearing capacity.

When the angle of the wedge with the horizontal surface $\omega$ (just below the base of the foundation) is taken as $\phi$, the coefficients of bearing capacity used in Terzaghi's bearing capacity formula are

$$N_q=\frac{1}{1-\sin\phi}e^{(1.5\pi-\phi)\tan\phi} \tag{10.11}$$

$$N_c=(N_q-1)\cot\phi \tag{10.12}$$

$$N_r\approx(N_q-1)\tan(1.4\phi) \tag{10.13}$$

when, $\omega=\frac{\pi}{4}+\frac{\phi}{2}$, it becomes

$$N_q=K_p\, e^{\pi\tan\phi} \tag{10.14}$$

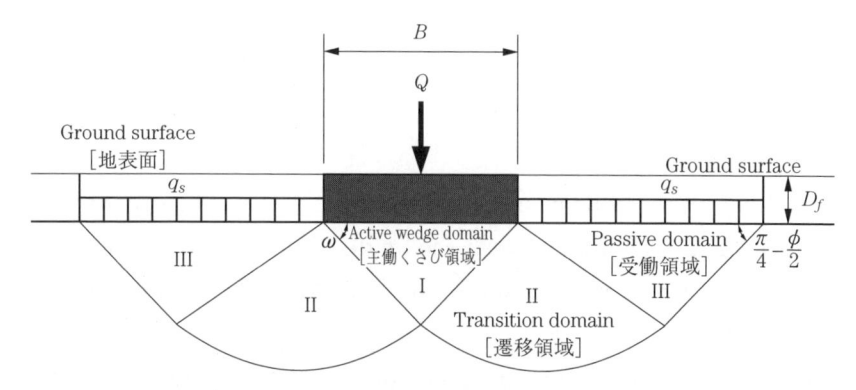

**Figure 10.5** Assumptions for Terzaghi's bearing capacity formula
**図 10.5** テルツァーギの支持力公式の仮定

大きな違いはない。

　ここでは，一般によく用いられるテルツァーギの支持力公式について考えてみる。テルツァーギの支持力公式は，全般せん断破壊を仮定し，**図 10.5** に示すように，塑性領域を，基礎直下の主働くさび領域Ⅰ，遷移領域Ⅱ，受働領域Ⅲに分け，Ⅰの領域と基礎が一体となって貫入することで，Ⅱの対数らせんを示す領域を介してⅢの受動領域を形成すると仮定することで，極限支持力を求めるものである。

　地盤に幅 $B$ の基礎が荷重 $Q$ で載荷するとともに，地盤表面に $q_s$ の分布荷重が載荷するときの極限支持力は，式(10.10)で表される。

$$\frac{Q}{B}(=q) = \frac{\gamma_t}{2} B N_r + c N_c + q_s N_q \tag{10.10}$$

$$= \frac{\gamma_t}{2} B N_r + c N_c + \gamma_{t2} D_f N_q \tag{10.10'}$$

　ここで，$q$：地盤の極限支持力度〔kN/m²〕，$\gamma_t$：地盤の単位体積重量〔kN/m³〕，$B$：基礎幅〔m〕，$C$：粘着力〔kN/m²〕，$q_s$：地盤表面の分布荷重（サーチャージ荷重）〔kN/m²〕，$N_r$, $N_c$, $N_q$：支持力係数，である。

　テルツァーギの支持力公式で用いられる支持力係数は，基礎直下のくさびが水平面となす角度 $\omega$ が $\phi$ の場合

$$N_c = (N_q - 1) \cot\phi \tag{10.15}$$
$$N_r \approx 2(N_q + 1) \tan\phi \tag{10.16}$$

When $\phi = 0$, $N_c$ is given by
$$N_c = 1.5\pi + 1 = 5.71 \tag{10.17}$$

Terzaghi's ultimate bearing capacity formula is used for continuous footings. Relationships between the angle of internal friction and the coefficients of bearing capacity for ($N_r$, $N_c$ and $N_q$) are given in **Table 10.1**.

Real foundations come in various shapes and sizes. A general bearing capacity formula that includes the shape (geometry) of a foundation

Table 10.1 Angle of internal friction and coefficients of bearing capacity ($\omega = \phi$)
表10.1 内部摩擦角と支持力係数

| $\phi$ | $N_c$ | $N_r$ | $N_q$ |
|---|---|---|---|
| 0 | 5.71 | 0.0 | 1.00 |
| 5 | 7.32 | 0.0 | 1.64 |
| 10 | 9.64 | 1.2 | 2.70 |
| 15 | 12.8 | 2.4 | 4.44 |
| 20 | 17.1 | 4.5 | 7.48 |
| 25 | 25.0 | 9.2 | 12.7 |
| 30 | 37.2 | 20.0 | 22.5 |
| 35 | 57.8 | 44.0 | 41.4 |
| 40 | 95.6 | 114.0 | 81.2 |
| 45 | 172.0 | 320.0 | 173.0 |

Table 10.2 Shape factors
表10.2 形状係数

| Shape functions [形状係数] | Continuous [連続] | Square shape [正方形] | Rectangular shape [長方形] | Circular shape [円形] |
|---|---|---|---|---|
| $\alpha$ | 1 | 1.3 | $1 + 0.3\dfrac{B}{L}$ | 1.3 |
| $\beta$ | 0.5 | 0.4 | $0.5 - 0.1\dfrac{B}{L}$ | 0.3 |

B : Shorter side of a rectangular foundation [長方形基礎の短辺]
L : Longer side of a rectangular foundation [長方形基礎の長辺]

$$N_q = \frac{1}{1-\sin\phi} e^{(1.5\pi-\phi)\tan\phi} \qquad (10.11)$$

$$N_c = (N_q - 1)\cot\phi \qquad (10.12)$$

$$N_r \approx (N_q - 1)\tan(1.4\phi) \qquad (10.13)$$

で求められる。

また，$\omega = \dfrac{\pi}{4} + \dfrac{\phi}{2}$ の場合

$$N_q = K_p\, e^{\pi\tan\phi} \qquad (10.14)$$

$$N_c = (N_q - 1)\cot\phi \qquad (10.15)$$

$$N_r \approx 2(N_q + 1)\tan\phi \qquad (10.16)$$

で求められる。

なお，$\omega = \phi$ において $\phi = 0°$ の場合の $N_c$ は

$$N_c = 1.5\pi + 1 = 5.71 \qquad (10.17)$$

で示される。

連続フーチングの設計においてテルツァーギの極限支持力公式が利用され，このときの $N_c$, $N_q$, $N_r$ における内部摩擦角と支持力係数との関係を**表 10.1** に示す。

基礎にはいろいろな形状があり，形状係数を導入した式(10.18)に示す一般化された支持力公式も提案されている。

$$q = \gamma_{t1} D_f N_q + \alpha c N_c + \beta \gamma_{t2} B N_r \qquad (10.18)$$

ここで，$\alpha$, $\beta$ は**表 10.2** に示す形状係数である。

**【例題 10.1】** 基礎の幅 $B = 5\,\mathrm{m}$，根入れ深さ $D_f = 1\,\mathrm{m}$ の連続フーチングを施工する場合の全般せん断破壊に対する極限支持力を求めよ（**図 10.6**）。なお，土の単位体積重量は $\gamma_t = 18\,\mathrm{kN/m^2}$，粘着力は $c = 10\,\mathrm{kN/m^2}$，内部摩擦角は $\phi = 30°$ である。

**解答** テルツァーギの極限支持力公式における支持力係数を求めると，表 10.1 より

$N_c = 37.2$, $N_r = 20.0$, $N_q = 22.5$

is given by Equation (10.18):
$$q = \gamma_{t1} D_f N_q + \alpha c N_c + \beta \gamma_{t2} B N_r \tag{10.18}$$
where $\alpha$, $\beta$ are the shape factors listed in **Table 10.2**.

**【Example 10.1】** Calculate the ultimate bearing capacity for a general shear failure of continuous footing of width ($B = 5$ m) and depth ($D_f = 1$ m) as shown in **Figure 10.6**. Take the unit weight of the soil as ($\gamma_t = 18$ kN/m²), the cohesion as ($c = 10$ kN/m²) and the angle of internal friction as ($\phi = 30°$).

**Solution** From Table 10.1, the bearing capacity coefficients for Terzaghi's ultimate bearing capacity formula are
$$N_c = 37.2, \ N_r = 20.0, \ N_q = 22.5$$
Taking $q_s$ as $\gamma_t \times D_f$, the ultimate bearing capacity is calculated by using Equation (10.10) as follows:
$$q = \frac{\gamma_t}{2} B N_r + c N_c + \gamma_t D_f N_q = \frac{18}{2} \times 5 \times 20 + 10 \times 37.2 + 18 \times 1 \times 22.5$$
$$= 1\ 677 \text{ kN/m}^2$$

## 10.3 Deep Foundation

A foundation is called a deep foundation when it is constructed at a depth that yields a large ratio of the depth of the foundation to its width (generally, $D_f/B > 5$). For a deep foundation, pilings are used to transmit loads from a structure into a strong soil layer when the structure cannot be supported by soft or weak ground near the surface. There are three methods of pile construction:

① End bearing piles : The base of an end bearing pile rests on a relatively firm soil and the load of the structure is transmitted through the pile into this firm soil.

② Friction piles : Friction piles transmit the load of the structure to the soil through the frictional resistance (skin friction or

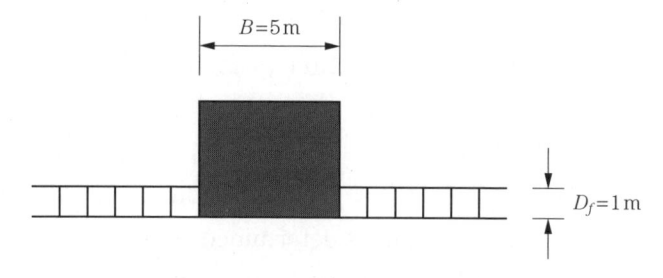

**Figure 10.6** A continuous footing
図 10.6 連続フーチング

である。
式 (10.10) の $q_s$ は $\gamma_t \times D_f$ であることを考慮し，極限支持力を求めると

$$q = \frac{\gamma_t}{2}BN_\gamma + cN_c + \gamma_t\,D_f\,N_q = \frac{18}{2} \times 5 \times 20 + 10 \times 37.2 + 18 \times 1 \times 22.5$$
$$= 1\,677\ \text{kN/m}^2$$

となる。

## 10.3　深 い 基 礎

　基礎幅と根入れ深さの比が大きく（一般には $D_f/B > 5$），深い位置に施工される基礎を深い基礎という。深い基礎には，軟弱地盤などの理由で構造物を直接支持することが困難な場合に，地盤内の硬い層まで荷重を伝えるために施工される杭などがある。

　深い基礎の支持力を求める場合，杭の機能を考慮して下記の三つについて考慮する必要がある。

① 先端支持杭：杭の先端支持力によって地盤内の硬い層に荷重を伝えるもの。
② 摩擦杭：杭周面に作用する摩擦抵抗によって荷重を支持するもの。
③ 締固め杭：振動などによって地盤を締固めることで地盤強度を増して荷重を支持するもの。

　深い基礎に対し，テルツァーギは単杭の極限支持力が，先端支持力 $Q_t$ と杭

cohesion) between the soil and the embedded surface of the pile.

③ Compaction piles : Compaction piles bear the loads from a structure by increasing the strength of the ground by either compaction or vibration.

For a deep foundation, Terzaghi found that the ultimate bearing capacity of a single pile is determined from the sum of the end bearing capacity $Q_t$ and the frictional resistance surrounding the pile $Q_s$ (**Figure 10.7**) and is given by

$$Q = Q_t + Q_s = q_t\, A_t + f_s\, A_s \tag{10.19}$$

where $q_t$ : Ultimate bearing capacity at the end of the pile (this is determined by calculating the ultimate bearing capacity of a shallow foundation), $A_t$ : Cross-sectional area at the end of the pile, $f_s$ : Frictional resistance of the surface area of the pile per unit area, $A_s$ : The surface area of the pile.

If the ground experiences large settlement after driving a pile, a downward frictional force develops on the pile. This downward frictional force is called the minus frictional force (or negative friction). The effect of negative friction is especially important for soft ground where large settlements surrounding the pile is observed. If groups of pilings are used, then the integrated effectiveness of the group of pilings must be assessed, especially for the pilings that have narrow pile-to-pile spacings.

【**Example 10.2**】 A 20 m pile with a diameter of 0.5 m was driven into the ground. Determine the ultimate bearing capacity of the pile using Terzaghi's ultimate bearing capacity formula. Assume that the unit weight of the soil is ($\gamma_t = 18$ kN/m²), the cohesion is ($c = 10$ kN/m²), the angle of internal friction is ($\phi = 30°$) and the skin friction of pile is ($f_s = 50$ kN/m²).

周面の摩擦抵抗力 $Q_s$ の和で求まるとして

$$Q = Q_t + Q_s = q_t A_t + f_s A_s \tag{10.19}$$

を示した（図 10.7）。

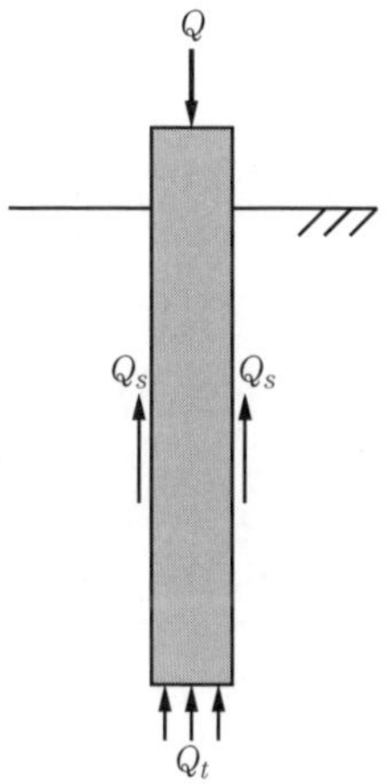

**Figure 10.7** Bearing capacity of a deep foundation
図 10.7 深い基礎の支持力

ここで，$q_t$：杭先端の極限支持力で，浅い基礎に対する極限支持力により求める，$A_t$：杭先端の断面積，$f_s$：単位面積あたりの杭の周面摩擦力，$A_s$：杭の周面積，である。

杭の施工後，周辺地盤の沈下が大きくなると，杭に下向きの摩擦力が発生することがあり，これを負の摩擦力（ネガティブフリクション）と呼ぶ。軟弱な地盤などで杭の周辺地盤に大きな沈下がみられる場合には，この影響について考慮する必要がある。また，杭の間隔を狭くして施工した場合，杭が一体化した群杭の効果が現れるため，群杭に対する検討が必要となる。

【例題 10.2】 地盤に直径 0.5 m の杭を 20 m 打ち込んだ。このときの杭の極限支持力をテルツァーギの極限支持力公式を基に求めよ。なお，地盤の単位体積重量は $\gamma_t = 18$ kN/m²，粘着力 $c = 10$ kN/m²，内部摩擦角 $\phi = 30°$，杭の周面摩擦 $f_s = 50$ kN/m² とする。

**Solution**  From Table 10.1 the bearing capacity coefficients for ($\phi=30°$) are $N_c=37.2$, $N_r=20.0$ and $N_q=22.5$. Also, from Table 10.2, shape functions for circular foundation are $\alpha=1.3$ and $\beta=0.3$.
Therefore, the ultimate bearing capacity using Terzaghi's ultimate bearing capacity Equation (10.19) is:

$$\begin{aligned}Q &= Q_t + Q_s = q_t\, A_t + f_s\, A_s \\ &= (\gamma_t\, D_f\, N_q + \alpha c N_c + \beta \gamma_t\, B N_r) A_t + f_s\, A_s \\ &= (18 \times 20 \times 22.5 + 1.3 \times 10 \times 37.2 + 0.3 \times 18 \times 0.5 \times 20) \times \pi \times 0.25^2 \\ &\quad + \pi \times 0.5 \times 20 \\ &= 1\,726.5 \text{ kN}\end{aligned}$$

**解答** $\phi=30°$ における支持力係数は，表 10.1 より $N_c=37.2$, $N_r=20.0$, $N_q=22.5$ である。また，円形基礎の場合の形状係数は，表 10.2 より $\alpha=1.3$, $\beta=0.3$ である。式(10.19)のテルツァーギの極限支持力公式を基に極限支持力を求めると

$$Q=Q_t+Q_s=q_t A_t+f_s A_s$$
$$=(\gamma_t D_f N_q+\alpha c N_c+\beta\gamma_t BN_\gamma)A_t+f_s A_s$$
$$=(18\times20\times22.5+1.3\times10\times37.2+0.3\times18\times0.5\times20)\times\pi\times0.25^2$$
$$\quad+\pi\times0.5\times20$$
$$=1\,726.5\text{ kN}$$

となる。

# Chapter 11   Slope Stability

## 11.1   Limit Equilibrium Method

Not all ground surfaces are flat like a sea or pond. Geographical features such as mountains and valleys affect the evenness of the ground. Such geographical features maintain their form because soils have shear strength while fluids (such as water) do not. The strength of a soil can also resist gravitational forces that tend to alter the shape or surface of the ground. However, a slope can collapses after significant rainfalls or an earthquake, etc. A slip surface is often observed at locations where slope failure (or collapse) has occurred as in **Figure 11.1**.

**Figure 11.1**   A picture of a slope failure showing the slip surface
図 11.1   斜面の崩壊の状況

# 11章 斜面安定

## 11.1 極限釣合い法

地面は平らな海や池とは違い，山や谷といった凸凹の地形をしている。山や谷などの地形がその形状を保てるのは，土には水などの流体には存在しないせん断強さが存在し，地球の重力によって生じる形を変化させようとする力に抵抗できるためである。しかし，安定している斜面も降雨や地震などによって斜面が崩壊することがある。斜面が崩壊した現場を見ると，**図 11.1** のようにすべり面（せん断面）を確認することができる。

斜面が崩壊する原因としては，すべり面上の体積力とすべり面で発揮されるせん断強さの釣合いが崩れることがあげられる。これを単純化すると，**図 11.2** に示すような，すべり土塊をブロック，すべり面を斜面と考えた斜面上にブロックを置いた剛体の釣合い問題と言い換えることができる。ここで，図

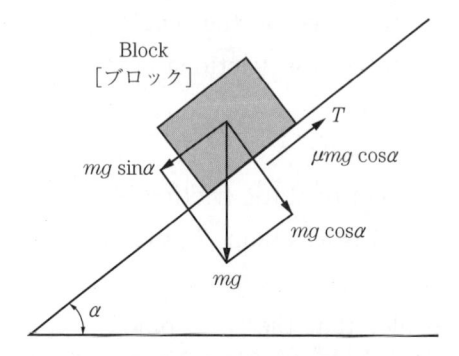

**Figure 11.2** Balance of forces for a solid block on a slope (inclined plane)
図 11.2 斜面上のブロックの釣合い

128　Chapter 11　Slope Stability

A breakdown in equilibrium or in the balance of forces between the weight of soil on the slip surface and the shear strength of the slip surface can lead to slope failure. This phenomenon can be understood from basic physics by using a solid block resting on an inclined plane (smooth slope) to represent the soil (**Figure 11.2**) sitting on a slip surface. If the block is at rest, then there is no net force on the block and it remains at rest on the inclined slope.

Using Newton's First Law, we can consider the stability of a solid block of weight $m$ at rest on a slope (inclined plane) of angle $\alpha$ as shown in Figure 11.2.

Equilibrium requires that there is no net force on the block. Using an $x$-$y$ coordinate system aligned with the inclined slope, the sum of forces that act on the block in the x direction ($\Sigma F_x = 0$) yields

$$T - mg \sin\alpha = 0 \tag{11.1}$$

$$T = mg \sin\alpha \tag{11.2}$$

where T is the static frictional force between the block and the slope. Summing the forces perpendicular to the slope ($y$-direction, $\Sigma F_y = 0$) results in

$$N - mg \cos\alpha = 0 \tag{11.3}$$

$$N = mg \cos\alpha \tag{11.4}$$

where $N$ is the normal force on the block from the inclined slope. When the block is stable, the frictional force, T, is related to the normal force $N$ by

$$T \leq \mu N \, (= \mu mg \cos\alpha) \tag{11.5}$$

where $\mu$ is the coefficient of static friction. As long as the component of the weight ($mg \cos\alpha$) is less than $T$, the block will remain at rest, that is, it will not slip.

Now, consider that the solid block is a soil mass and the slope (inclined plane) is a slip surface (**Figure 11.3**) to study the stability of a half-infinite layer of ground with a ground angle of $\beta$,

## 11.1 Limit Equilibrium Method

11.2に示す角度 $\alpha$ の斜面上にある質量 $m$ のブロックの安定を考えてみる。

斜面と平行な方向の力の釣合いは

$$T - mg \sin\alpha = 0 \tag{11.1}$$

$$T = mg \sin\alpha \tag{11.2}$$

また，斜面と垂直な方向の力の釣合いは

$$N - mg \cos\alpha = 0 \tag{11.3}$$

$$N = mg \cos\alpha \tag{11.4}$$

である。

ここで，物体と斜面の間に摩擦法則が成り立ち，静止摩擦係数を $\mu$ とすると

$$T \leq \mu N \,(= \mu mg \cos\alpha) \tag{11.5}$$

のとき，ブロックは静止し安定している。

つぎに，ブロックをすべり土塊，斜面をすべり面と考え，図 11.3 に示すような地表勾配 $\beta$，地表厚さ $H$，土の単位体積重量 $\gamma_t$ の，半無限斜面におけるすべり（平行すべり）に対する安定を考えてみる。

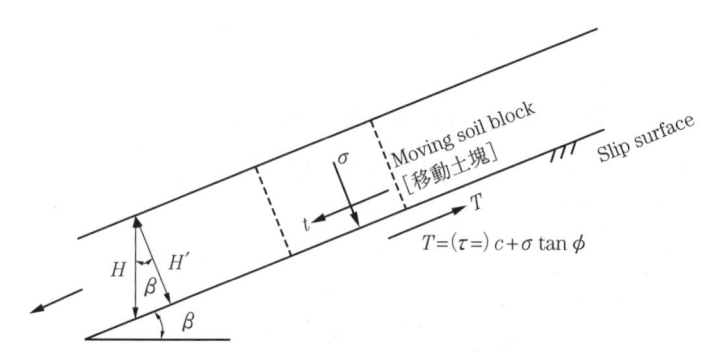

**Figure 11.3** Stability of a half-infinite slope
図 11.3 半無限斜面における安定性の検討

すべり土塊により発生する垂直応力 $\sigma$，せん断応力 $t$ は，それぞれ

$$\sigma = \gamma_t H' \cos\beta = \gamma_t (H \cos\beta) \cos\beta = \gamma_t H \cos^2\beta \tag{11.6}$$

$$t = \gamma_t H' \sin\beta = \gamma_t (H \cos\beta) \sin\beta = \gamma_t H \sin\beta \cos\beta \tag{11.7}$$

である。

a ground thickness of $H$ and a unit weight of soil $\gamma_t$.
The normal stress $\sigma$ and shear stress $t$ acting on the slip surface are expressed as follows,

$$\sigma = \gamma_t H' \cos\beta = \gamma_t (H \cos\beta)\cos\beta = \gamma_t H \cos^2\beta \qquad (11.6)$$

$$t = \gamma_t H' \sin\beta = \gamma_t (H \cos\beta)\sin\beta = \gamma_t H \sin\beta \cos\beta \qquad (11.7)$$

Now, compare the existing shear strength of soil $T$ to shear strength that develops on the slip surface $t$.
When $t \leq T$, the slope is stable.
The safety factor $F_s$ of the slope is calculated by using the ratio of the shear strength of the slip surface $t$ to the shear strength of the soil, $T$. The expression for the safety factor is

$$F_s = \frac{T}{t} \qquad (11.8)$$

The slope is stable when $F_s$ is greater than 1 (one) because it satisfies the condition that $t \leq T$.
The shear strength of soil can be calculated by using Coulomb's failure criterion as follows

$$\tau = c + \sigma \tan\phi \qquad (11.9)$$

The shear strength $T$ developed in the soil is obtained by substituting Equation (11.6) into Equation (11.9) yielding:

$$T = c + \sigma \tan\phi = c + \gamma_t H \cos^2\beta \tan\phi \qquad (11.10)$$

Now, the factor of safety $F_s$ is found by substituting Equations (11.7) and (11.10) into Equation (11.8):

$$F_s = \frac{c + \gamma_t H \cos^2\beta \tan\phi}{\gamma_t H \sin\beta \cos\beta} \qquad (11.11)$$

For a sandy ground where there is no cohesion ($c = 0 \text{ kN/m}^2$), the safety factor becomes,

$$F_s = \frac{\gamma_t H \cos^2\beta \tan\phi}{\gamma_t H \sin\beta \cos\beta} = \frac{\cos\beta \tan\phi}{\sin\beta} = \frac{\tan\phi}{\tan\beta} \qquad (11.12)$$

When the half infinite slope is in the saturated condition, the water pressure $u$ developed in the ground is given by

ここで，土のもっているせん断強さ $T$ とすべり土塊により発生するせん断応力 $t$ とを比較し，$t \leqq T$ の場合に斜面は安定すると考えることができる。

ここで，$T$ が $t$ の何倍耐えることができるかを考え

$$F_s = \frac{T}{t} \tag{11.8}$$

を用いて斜面の安定性を評価してみる。

このとき，$F_s$ は安全率と呼ばれ，$F_s$ が1以上の場合 $t \leqq T$ を満足できるため，斜面は安定であると考えることができる。

ところで，土のせん断強度はクーロンの破壊基準

$$\tau = c + \sigma \tan\phi \tag{11.9}$$

で求められることから，
式(11.9)に式(11.6)を代入し，土が発揮できるせん断強度 $T$ を求めると

$$T = c + \sigma \tan\phi = c + \gamma_t H \cos^2\beta \tan\phi \tag{11.10}$$

となる。

ここで，式(11.8)に式(11.7)と式(11.10)を代入し安全率 $F_s$ を求めると

$$F_s = \frac{c + \gamma_t H \cos^2\beta \tan\phi}{\gamma_t H \sin\beta \cos\beta} \tag{11.11}$$

となる。

このとき，粘着力 $c = 0 \, \text{kN/m}^2$ の砂地盤の場合，式(11.11)は

$$F_s = \frac{\gamma_t H \cos^2\beta \tan\phi}{\gamma_t H \sin\beta \cos\beta} = \frac{\cos\beta \tan\phi}{\sin\beta} = \frac{\tan\phi}{\tan\beta} \tag{11.12}$$

で示される。

ここで，半無限斜面が飽和状態にある場合，地盤内で発生する間隙水圧 $u$ は，水の単位体積重量を $\gamma_w$ とすると

$$u = \gamma_w H \cos^2\beta \tag{11.13}$$

であるため，
有効応力 $\sigma' = \sigma - u$ を考えると，垂直応力 $\sigma'$ は

$$\sigma' = \gamma_t H \cos^2\beta - \gamma_w H \cos^2\beta = (\gamma_t - \gamma_w) H \cos^2\beta \tag{11.14}$$

となる。

$$u = \gamma_w H \cos^2 \beta \tag{11.13}$$

where, $\gamma_w$ is unit weight of water.

Considering the effective stress ($\sigma' = \sigma - u$), the normal stress $\sigma'$ is given by

$$\sigma' = \gamma_t H \cos^2 \beta - \gamma_w H \cos^2 \beta = (\gamma_t - \gamma_w) H \cos^2 \beta \tag{11.14}$$

Therefore, the safety factor of a half infinite slope in the saturated condition is

$$F_s = \frac{c' + (\gamma_t - \gamma_w) H \cos^2 \beta \tan \phi'}{\gamma_t H \sin \beta \cos \beta} \tag{11.15}$$

For a sandy ground where the cohesion is zero ($c' = 0 \text{ N/m}^2$), the safety factor is

$$F_s = \frac{\gamma_t - \gamma_w}{\gamma_t} \frac{\tan \phi}{\tan \beta} \tag{11.16}$$

**【Example 11.1】** A vertical slope of clayey soil develops a slip surface with an angle of $\alpha$ relative to the horizontal as shown in **Figure 11.4**. The angle of internal friction is (0°). Calculate the safety factor for this slope.

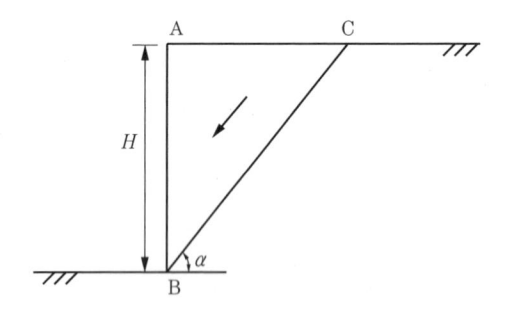

**Figure 11.4** Stability of a vertical slope
図 11.4 垂直斜面の安定性

**Solution** From **Figure 11.5.**, the weight $W$ of the soil wedge (ABC) bounded by the slip surface is

したがって，飽和状態にある半無限斜面の安全率は

$$F_s = \frac{c' + (\gamma_t - \gamma_w) H \cos^2\beta \tan\phi'}{\gamma_t H \sin\beta \cos\beta} \tag{11.15}$$

で示される。

また，$c' = 0 \text{ kN/m}^2$ の砂地盤の場合，式(11.15)は

$$F_s = \frac{\gamma_t - \gamma_w}{\gamma_t} \frac{\tan\phi}{\tan\beta} \tag{11.16}$$

となる。

**【例題 11.1】** 図 11.4 に示す内部摩擦角が $0°$ である粘土の垂直斜面で，水平面と $\alpha$ の角度ですべりが生じるときの安全率を求めよ。

**解答** 図 11.5 に示すように，すべり面で囲まれた土塊 ABC の重量 $W$ は，土の単位体積重量を $\gamma_t$ とすると

$$W = \frac{1}{2} H \frac{H}{\tan\alpha} \gamma_t$$

より，土塊 ABC がすべり面に沿って移動する力 $F$ は

$$F = W \sin\alpha = \frac{1}{2} H^2 \gamma_t \frac{\sin\alpha}{\tan\alpha}$$

となる。

内部摩擦角が $0°$ の粘土地盤において，すべり面における粘着力 $c$ のすべりに対す

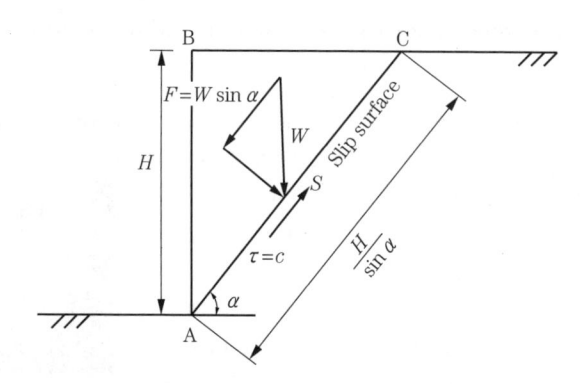

**Figure 11.5** Schematic for the calculation of the safety factor for a vertical slope
図 11.5 垂直斜面の安定性の計算

$$W = \frac{1}{2} H \frac{H}{\tan\alpha} \gamma_t$$

where $\gamma_t$ is the unit weight of the soil.
The force needed to cause the soil wedge (ABC) to slip along the shear surface is given by

$$F = W \sin\alpha = \frac{1}{2} H^2 \gamma_t \frac{\sin\alpha}{\tan\alpha}$$

For a clayey ground with angle of internal friction of (0°), the resisting force $S$ for the entire slip surface due to cohesion $c$ of the slip surface is

$$S = \frac{H}{\sin\alpha} c$$

Therefore, the safety factor $F_s$ is

$$F_s = \frac{S}{F} = \frac{\dfrac{cH}{\sin\alpha}}{\dfrac{1}{2} H^2 \gamma_t \dfrac{\sin\alpha}{\tan\alpha}} = \frac{2cH}{H^2 \gamma_t \sin\alpha \cos\alpha}$$

$$= \frac{4c}{\gamma_t H \sin 2\alpha}$$

---

### Adhémar Jean Claude Barré de Saint-Venant
(サンブナン,1797〜1886)

フランスの土木工学者。エコール・ポリテクニークで学ぶ。サンブナンは主として機械工学,弾性学,流体力学の分野を教えた。

最近の構造力学の教科書ではあまりお目にかからないが,サンブナンの原理(Saint-Venant's principle)というものがある。これは,固定条件の影響する範囲は,だいたい固定辺より固定長に等しい長さの部分であり,それ以下の応力分布には固定方法は影響しない。これは引張り,圧縮,曲げその他の問題についても成立する関係である。

図17 サンブナン
(出典:St-Andrews 大学,スコットランド)

る抵抗力 $S$ は，すべり面全体では

$$S = \frac{H}{\sin\alpha} c$$

となる。

したがって，安全率 $F_s$ は

$$F_s = \frac{S}{F} = \frac{\dfrac{cH}{\sin\alpha}}{\dfrac{1}{2} H^2 \gamma_t \dfrac{\sin\alpha}{\tan\alpha}} = \frac{2cH}{H^2 \gamma_t \sin\alpha \cos\alpha}$$

$$= \frac{4c}{\gamma_t H \sin 2\alpha}$$

である。

---

### Jules Dupuit（デュピュイ，1804〜1866）

フランスの土木工学者，経済学者。ナポレオン・ボナパルトの支配下にあったイタリア・フォッサーノ生まれ。エコール・ポリテクニークで土木工学を学び，1843年フランスの道路システムに関してレジオンドヌール勲章を受賞。地下水に関するデュピュイの公式で知られる。

$$Q = \pi k \frac{H^2 - h_0^2}{\log_e \dfrac{R}{r_0}}$$

図18 デュピュイ

$R$ は，それより遠方では地下水位低下は起こらないという半径で，影響圏半径という。通常，井戸半径 $r_0$ の 3 000〜5 000 倍，または 500〜1 000 m をとることが多い。

## 11.2 Stability Analysis Considering Circular Slip Surface

In practice, a sliding-soil wedge is divided into a number of slices when performing slope stability analyses. A slice method is used that considers the equilibrium condition for each slice. The first step in stability analysis when using the slice method is to determine the shape of the slip surface. From practical observations, a slip surface may be a straight line or a circular arc or a combination of these. For simplicity, it is taken to be circular during analysis. Now, consider a simplified method (Fellenius method) for stability analysis using a circular slip surface.

  The simplified method divides the sliding-soil wedge into several vertical slices. The resisting moment from the shear strength of the slip surface is calculated with respect to the center of the circular arc. The ratio of the resisting moment to the sliding moment (based on the sliding-soil wedge) is taken as the safety factor ($F_s$ = Resisting moment/Sliding moment). In this simplified method, the horizontal and vertical forces between the slices are assumed to be equal. Though this assumption is contradictory to reality, this assumption is widely used currently because of its conveniences for calculations.

  Now, let's consider the balance of forces for one of the slices as shown in **Figure 11.6**.

Because the simplified method assumes that inter-slice forces are equal and opposite, ((b)-①) we get

$$H_n - H_{n+1} = 0 \qquad (11.17)$$

$$V_n - V_{n+1} = 0 \qquad (11.18)$$

The sliding moment of a slice about the center of the circular arc is given by

## 11.2 円弧すべり面の安定解析

実際の斜面における安定性の評価を行う場合,すべり土塊をいくつかのスライスに分割し,個々の分割片ごとに土の釣合いを考える分割法が用いられる。この分割法により安定解析を行う場合,すべり面で囲まれたすべり土塊の形状を前もって仮定しておくことが必要となる。すべり面の形状は,実際の観察などから直線,円弧などが組み合わさったものであると考えられるが,解析上の簡便さから円弧として近似することが行われる。ここで円弧すべりを考えた簡便法(フェレニウス法)による安定解析について考えてみる。

簡便法は,すべり土塊を数個以上の垂直なスライスに分割し,すべり円弧の中心に関して,すべり面で発揮されるせん断強さによる抵抗モーメントと,すべり土塊による滑動モーメントの比($F_s$=抵抗モーメント/滑動モーメント)を安全率として,斜面の安定性を評価する手法である。簡便法は,分割されたスライス間での水平方向,垂直方向の断面力が等しいと仮定することで求められており,力学的矛盾を含んでいるが,計算が容易であるため現在まで広く用

(a) The slice method
[分割法]

(b) Inter-slice forces of a sliding-soil wedge
[土塊のスライス間に働く力]

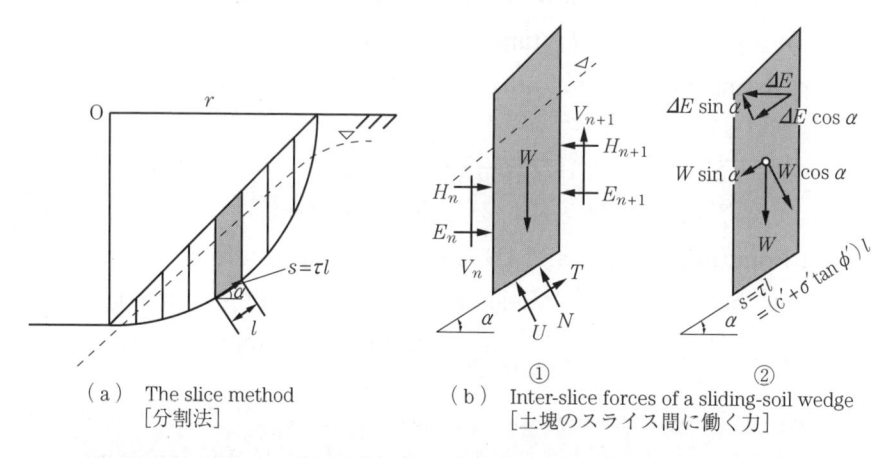

**Figure 11.6** The slice method and inter-slice forces of a sliding-soil wedge
図 11.6 分割法と土塊のスライス間に働く力

$$m_D = rW\sin\alpha + r\Delta E \cos\alpha \tag{11.19}$$

where, $r$ is the radius of the circular arc and ($\Delta E = E_{n+1} - E_n$) is the difference in water pressure between the slices ((b)-②). Therefore, the sliding moment for all of the slices of a sliding-soil wedge $M_D$ is given by

$$M_D = \Sigma(rW\sin\alpha_i + r\Delta E \cos\alpha_i) \tag{11.20}$$

By using Coulomb's failure criterion ($\tau = c' + \sigma'\tan\phi'$) and considering the effective stress ($\sigma' = (N-U)/l$), the resisting shear strength developed at the slip surface of a slice is

$$\tau = c' + \frac{\tan\phi'(N-U)}{l} \tag{11.21}$$

Taking the length of each element is $l$, Equation (11.21) becomes

$$s = \tau l = c'l + (N-U)\tan\phi' \tag{11.22}$$

where,

$$N = W\cos\alpha - \Delta E \sin\alpha \tag{11.23}$$
$$U = ul \tag{11.24}$$

here, $u$ is the interslice pore-water pressure and $l$ is the length of slip surface, Multiplying Equation (11.22) by the radius of a circular arc, the resisting moment of a slice about the center of circular arc is

$$m_R = r\{c'l + (N-U)\tan\phi'\} \tag{11.25}$$

The resisting moment for all of the slices of the sliding-soil wedge $M_R$ is given by

$$\begin{aligned}M_R &= \Sigma r\{c'l + (N-U)\tan\phi'\} \\ &= \Sigma r\{c'l + (W\cos\alpha - \Delta E \sin\alpha - ul)\tan\phi'\}\end{aligned} \tag{11.26}$$

From Equations (11.20) and (11.21), the safety factor $F_s$ determined by using the relation ($F_s$ = Resisting moment/Sliding moment) is

$$F_s = \frac{M_R}{M_D} = \frac{\Sigma\{c'l + (W\cos\alpha - \Delta E \sin\alpha - ul)\tan\phi'\}}{\Sigma(W\sin\alpha + \Delta E \cos\alpha)} \tag{11.27}$$

If ($\Delta E = 0$) is used in Equation (11.27), this is called the simplified method. The safety factor for the simplified method is given by

## 11.2 Stability Analysis Considering Circular Slip Surface

いられている。

ここで，**図 11.6** に示すようなすべり土塊をいくつかに分割した中の，一つのスライスについて考えてみる。

簡便法は，スライス間力を等しいと仮定しているため

$$H_n - H_{n+1} = 0 \tag{11.17}$$

$$V_n - V_{n+1} = 0 \tag{11.18}$$

が成り立つと仮定する（(b)-①）。

すべり円弧の半径を $r$ とすると，すべり円弧の中心に関し一つのスライスに対する滑動モーメントは，スライス間の水圧の差を $\Delta E = E_{n+1} - E_n$ とすると

$$m_D = rW \sin\alpha + r \Delta E \cos\alpha \tag{11.19}$$

となる（(b)-②）。

したがって，すべり土塊全体の全スライスに対する滑動モーメント $M_D$ は

$$M_D = \sum(rW \sin\alpha_i + r \Delta E \cos\alpha_i) \tag{11.20}$$

で示される。

一つのスライスにおいてすべり面で発揮されるせん断抵抗力 $s$ は，クーロンの破壊基準 $\tau = c' + \sigma' \tan\phi'$ において，有効応力 $\sigma' = (N-U)/l$ を考慮すると

$$\tau = c' + \frac{\tan\phi'(N-U)}{l} \tag{11.21}$$

となり，各要素すべり面の長さ $l$ を考慮すると

$$s = \tau l = c'l + (N-U)\tan\phi' \tag{11.22}$$

で示される。

ここで

$$N = W\cos\alpha - \Delta E \sin\alpha \tag{11.23}$$

$$U = ul \tag{11.24}$$

ここで，$u$：スライス間に働く間隙水圧，$l$：すべり面の長さである。

すべり円弧の中心に関する一つのスライスに対する抵抗モーメントは，式

$$F_s = \frac{\Sigma\{c'l + (W\cos\alpha - ul)\tan\phi'\}}{\Sigma W \sin\alpha} \tag{11.28}$$

Bishop assumed that the inter-slice forces in the vertical direction were also in equilibrium. He proposed the following equation for stability analysis considering a sliding circular arc against an over condition, making the problem statically indeterminate by one term.

$$F_s = \frac{\Sigma\{c'l\cos\alpha + (W - ul\cos\alpha)\tan\phi'\}/m}{\Sigma W \sin\alpha} \tag{11.29}$$

$$m = \cos\alpha\left(1 + \frac{\tan\alpha\tan\phi'}{F_s}\right) \tag{11.30}$$

As can be seen in above equations, $F_s$ appears on both the sides of the equation in Bishop's stability analysis. Therefore, an iterative approach is used for the calculations and continues until convergence of $F_s$ is reached.

**【Example 11.2】** Calculate the safety factor $F_s$ for a (45°) slope as shown in **Figure 11.7** by using the slice method and assuming a sliding-circular arc. Take the unit weight of soil to be ($\gamma_t = 18$ kN/m³), the cohesion to be (10 kN/m²) and the angle of internal friction to be (20°).

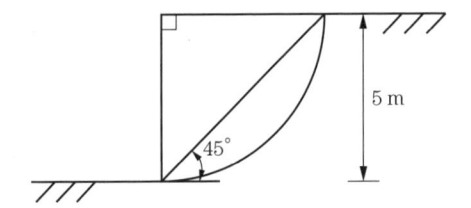

**Figure 11.7** Evaluation of slope stability considering sliding-circular arc
図 11.7　斜面の円弧すべりに対する安定性の評価

**Solution**　　Here, the slope is divided into 6 slices as shown in **Figure 11.8**.

(11.22)に円弧の半径を掛け

$$m_R = r\{c'l + (N-U)\tan\phi'\} \tag{11.25}$$

となる。

すべり土塊全体の全スライスに対する抵抗モーメント $M_R$ は

$$\begin{aligned} M_R &= \sum r\{c'l + (N-U)\tan\phi'\} \\ &= \sum r\{c'l + (W\cos\alpha - \Delta E\sin\alpha - ul)\tan\phi'\} \end{aligned} \tag{11.26}$$

で示される。

式(11.20),式(11.26)より安全率 $F_s$ は,$F_s$=滑動モーメント/抵抗モーメントより

$$F_s = \frac{M_R}{M_D} = \frac{\sum\{c'l + (W\cos\alpha - \Delta E\sin\alpha - ul)\tan\phi'\}}{\sum(W\sin\alpha + \Delta E\cos\alpha)} \tag{11.27}$$

となる。

式(11.27)において,$\Delta E = 0$ とおいたものを簡便法と呼び,このときの安全率は

$$F_s = \frac{\sum\{c'l + (W\cos\alpha - ul)\tan\phi'\}}{\sum W\sin\alpha} \tag{11.28}$$

で示される。

ビショップは,スライス間力のうち垂直方向の断面力が釣り合うと仮定することで,不静定次数が一つ条件過多となる円弧すべりに対する次式の安定解析を提案している。

$$F_s = \frac{\sum\{c'l\cos\alpha + (W - ul\cos\alpha)\tan\phi'\}/m}{\sum W\sin\alpha} \tag{11.29}$$

$$m = \cos\alpha\left(1 + \frac{\tan\alpha\tan\phi'}{F_s}\right) \tag{11.30}$$

ビショップの安定解析は,上式のように両辺に $F_s$ が入っているため,$F_s$ の収束解を求めるため反復計算を行う必要がある。

【例題 11.2】 45°の斜面における図 11.7 に示すような円弧すべりに対する安全率を,分割法によって求めよ。なお,この斜面の単位体積重量は $\gamma_t = 18$ kN/m³,粘着力 10 kN/m²,内部摩擦角 20° とする。

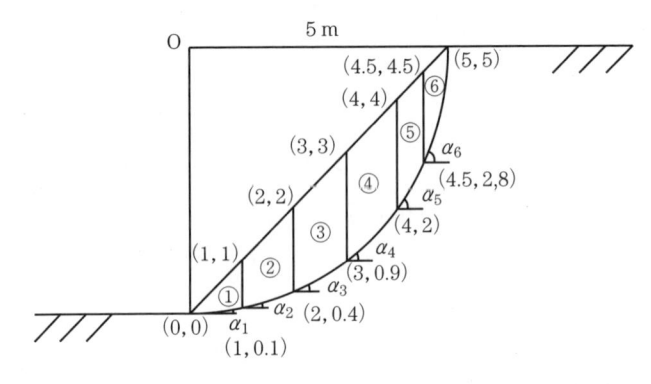

**Figure 11.8** A sliding-circular arc and the slice method
図 11.8　円弧すべりと分割法

The length of the slip surface $L$ is $L = \dfrac{10\pi}{4} = 7.85$ m

The safety factor is calculated according to **Table 11.1** and by using Equation (11.27).

Considering the total stress, the safety factor is

$$F_s = \frac{M_R}{M_D} = \frac{\Sigma\{cl + (W\cos\alpha\tan\phi)\}}{\Sigma(W\sin\alpha)}$$

yielding

$$F_s = \frac{10 \times 7.85 + 38.33}{79.12} = 1.48$$

## 11.2 Stability Analysis Considering Circular Slip Surface

**解答** ここでは，図 11.8 に示す六つの要素にスライスする。

すべり面の長さ $L$ は，$L = \dfrac{10\pi}{4} = 7.85$ m であり，表 11.1 に従って式 (11.27) を基に安全率を求める。

全応力における安全率は

$$F_s = \frac{M_R}{M_D} = \frac{\Sigma\{cl + (W\cos\alpha\tan\phi)\}}{\Sigma(W\sin\alpha)}$$

より

$$F_s = \frac{10 \times 7.85 + 38.33}{79.12} = 1.48$$

となる。

**Table 11.1** Results of the slice method
**表 11.1** 分割法の計算結果

| Slice<br>[要素]<br>No. | Angle<br>[角度]<br>$a[°]$ | Cross-sectional<br>area [断面積]<br>$A[\text{m}^2]$ | Self weight<br>[自重]<br>$W[\text{kN}]$ | $W\sin\alpha$ | $W\cos\alpha$ | $W\cos\alpha\tan\phi$ |
|---|---|---|---|---|---|---|
| 1 | 5.7 | 0.45 | 9.0 | 0.89 | 8.96 | 3.26 |
| 2 | 16.7 | 1.25 | 25.0 | 7.18 | 23.95 | 8.72 |
| 3 | 26.6 | 1.85 | 37.0 | 16.57 | 33.08 | 12.04 |
| 4 | 47.7 | 2.05 | 41.0 | 30.32 | 27.59 | 10.04 |
| 5 | 58.0 | 0.93 | 18.6 | 15.77 | 9.86 | 3.59 |
| 6 | 77.2 | 0.43 | 8.6 | 8.39 | 1.86 | 0.68 |
| 合計 | — | — | — | 79.12 | — | 38.33 |

# 付　録

## A．三軸圧縮試験の種類

| 試験の種類 | 適用土質 | 排水コックの開閉 | | 間隙水圧の測定 | 軸ひずみ速度：$\varepsilon_a$ | 求まる強度定数 | 試験結果の利用 |
|---|---|---|---|---|---|---|---|
| | | 圧密過程 | せん断過程 | | | | |
| 非圧密・非排水試験（UU） | 飽和粘性土 | 閉じる | 閉じる | しない | 1 %/min | $\phi_u$, $c_u$ | 非排水せん断強さ 粘性土地盤の短期安定問題 |
| 圧密・非排水試験（CU） | 飽和粘性土 | 開ける | 閉じる | しない | 1 %/min | $\phi_{cu}$, $c_{cu}$ | 圧密後の短期安定問題 |
| 圧密・非排水試験（CU） | | | | する | シルト分が多い，0.1 %/min 粘土分が多い，0.05 %/min | $\phi'$, $c'$ ($\phi_{cu}$, $c_{cu}$) | 有効応力解析 |
| 圧密・排水試験（CD） | 飽和土 | 開ける | 開ける | しない | $\varepsilon_a = \varepsilon_{af}/15t_c$ で 0.5 %/min を超えない $\varepsilon_{af}$：主応力差最大時の軸ひずみ $t_c$：圧密時間(min) | $\phi_d$, $c_d$ | 砂質土の安定問題 盛土の緩速施行 粘性土地盤の長期安定問題 |

付図1　三軸圧縮試験装置

## B. 土質試験の種類

| | 試験の目的 | 土質試験名 | 求められる値 | JIS, JGS |
|---|---|---|---|---|
| 土の物理的性質を求める試験 | 土(の三相)の状態 | 土の含水比試験 | 含水比：$w$ | JIS A 1203 |
| | | 土粒子の密度試験 | 土粒子密度：$\rho_s$ | JIS A 1202 |
| | | 土の湿潤密度試験 | 湿潤密度：$\rho_t$<br>乾燥密度：$\rho_d$ | JIS A 1225 |
| | | 砂の最小密度・最大密度試験 | 最小乾燥密度：$\rho_{d\,min}$<br>最大乾燥密度：$\rho_{d\,max}$ | JIS A 1224 |
| | | 土の保水試験 | 含水比：$w$<br>ポテンシャル：$\psi$ | JGS 0151 |
| | 土の工学的分類 | 土の粒度試験 | (粒径加積曲線)<br>有効径：$D_{10}$<br>平均粒径：$D_{50}$<br>均等係数：$U_c$<br>曲率係数：$U_c'$ | JIS A 1204 |
| | | 土の液性限界・塑性限界試験 | 液性限界：$w_L$<br>塑性限界：$w_P$<br>塑性指数：$I_P$ | JIS A 1205 |
| 土の力学的性質を求める試験 | 土の締固め | 突固めによる土の締固め試験 | (締固め曲線)<br>最適含水比：$w_{opt}$<br>最大乾燥密度：$\rho_{d\,max}$ | JIS A 1210 |
| | | CBR試験 | CBR値 | JIS A 1211 |
| | 土の圧縮性 | 土の圧密試験 | (圧縮曲線)<br>圧密降伏応力：$p_c$<br>圧縮指数：$C_c$<br>体積圧縮係数：$m_v$<br>圧密係数：$c_v$ | JIS A 1217 |
| | 土の強度・変形 | 土の一面せん断試験 | 内部摩擦角：$\phi$<br>($\phi'$, $\phi_{cu}$, $\phi_d$)<br>粘着力：$c$<br>($c'$, $c_{cu}$, $c_d$) | JGS 0560, 0561 |
| | | 土の三軸圧縮試験 | 内部摩擦角：$\phi$<br>($\phi'$, $\phi_u$, $\phi_{cu}$, $\phi_d$)<br>粘着力：$c$<br>($c'$, $c_u$, $c_{cu}$, $c_d$) | JGS 0520〜0524 |
| | | 土の一軸圧縮試験 | 非排水せん断強さ：$s_u$<br>一軸圧縮強さ：$q_u$<br>鋭敏比：$S_t$<br>変形係数：$E_{50}$ | JIS A 1216 |
| | 土の透水性 | 土の透水試験 | 透水係数：$k$ | JIS A 1218 |
| 土の化学的性質を求める試験 | 土の化学的性質 | 土の強熱減量試験 | 強熱減量：$L_i$ | JIS A 1226 |
| | | 土懸濁液のpH試験 | pH | JGS 0211 |
| | | 土懸濁液の電気伝導率試験 | 電気伝導率：$\chi$ | JGS 0212 |

## C. ボーリング調査

付図2に示すボーリング機械を用いて地盤に孔をあけることをボーリングという。一般にボーリングされる孔径は 66 mm, 86 mm, 116 mm の3種類がある。ボーリング調査は，地盤状況を確認するために行われるもので，ボーリングの際，地表から到達点までの地盤をまるごと取り込むボーリングコア（付図3）の採取を行う場合がある。また，ボーリングの際，標準貫入試験により所定の深さごとの $N$ 値を求めるとともに，同時に土試料を採取し必要に応じて物理試験を実施する。ボーリング調査の結果は，ボーリングコアの観察による地盤の性状，地下水位状況，$N$ 値などを示したボーリング柱状図（付図4）としてまとめられる。

付図2　ボーリング機械 　　付図3　ボーリングコア

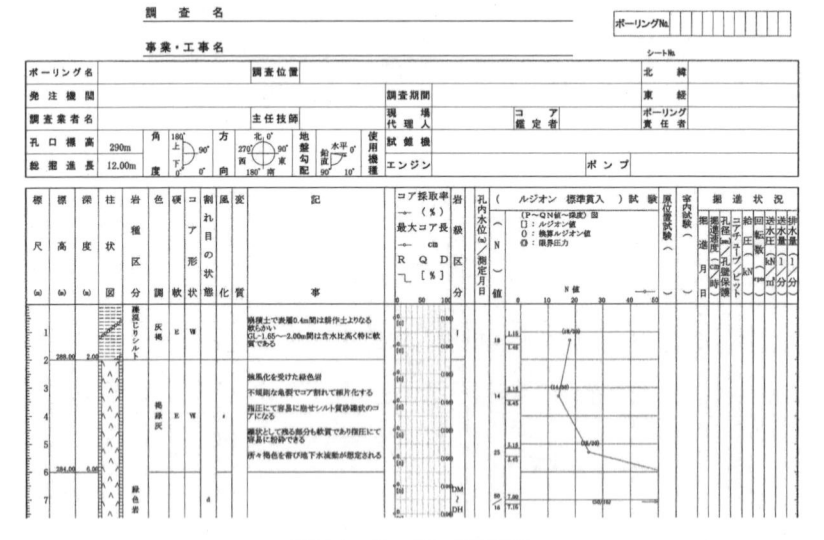

付図4　ボーリング柱状図

付録　147

## D. 日本地質概要図

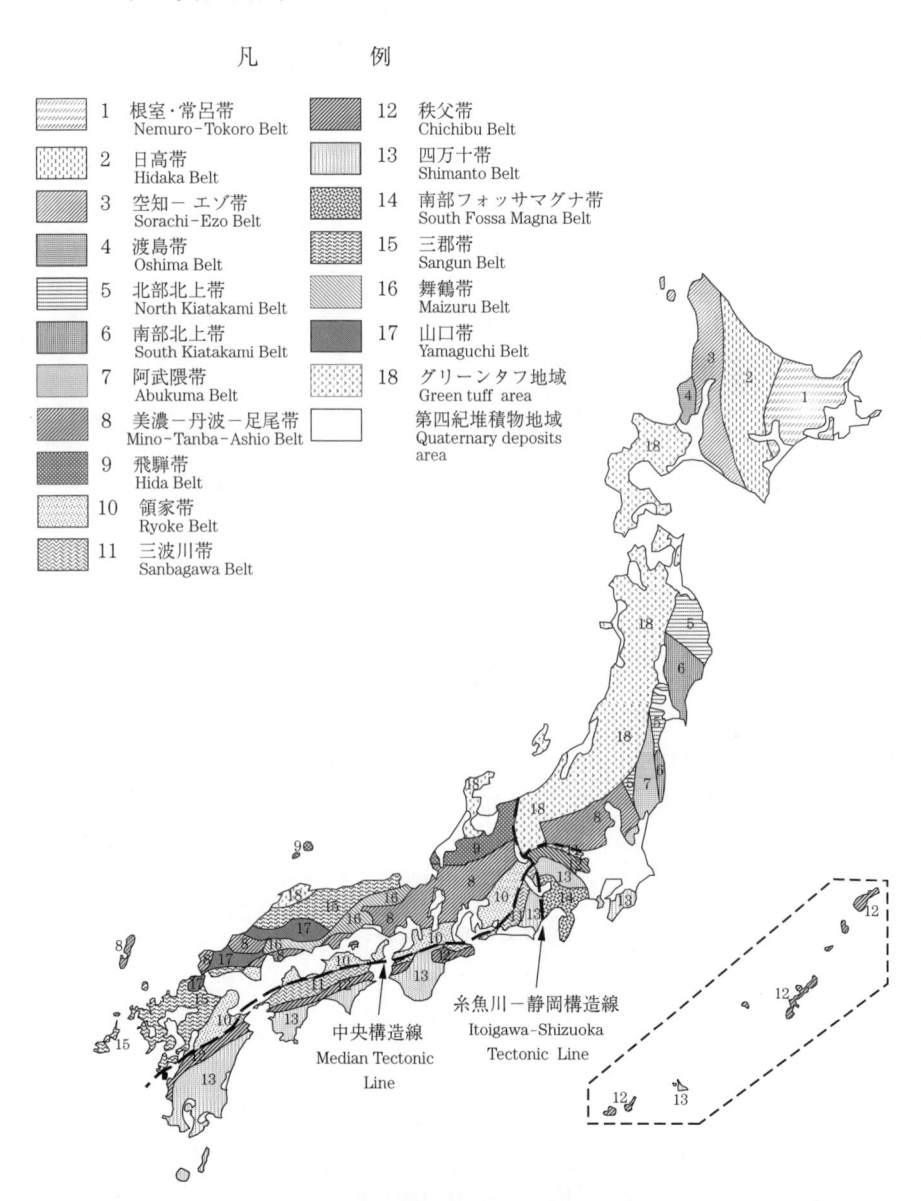

凡　例

| | | | | |
|---|---|---|---|---|
| | 1 | 根室・常呂帯 Nemuro-Tokoro Belt | 12 | 秩父帯 Chichibu Belt |
| | 2 | 日高帯 Hidaka Belt | 13 | 四万十帯 Shimanto Belt |
| | 3 | 空知ーエゾ帯 Sorachi-Ezo Belt | 14 | 南部フォッサマグナ帯 South Fossa Magna Belt |
| | 4 | 渡島帯 Oshima Belt | 15 | 三郡帯 Sangun Belt |
| | 5 | 北部北上帯 North Kiatakami Belt | 16 | 舞鶴帯 Maizuru Belt |
| | 6 | 南部北上帯 South Kiatakami Belt | 17 | 山口帯 Yamaguchi Belt |
| | 7 | 阿武隈帯 Abukuma Belt | 18 | グリーンタフ地域 Green tuff area |
| | 8 | 美濃ー丹波ー足尾帯 Mino-Tanba-Ashio Belt | | 第四紀堆積物地域 Quaternary deposits area |
| | 9 | 飛騨帯 Hida Belt | | |
| | 10 | 領家帯 Ryoke Belt | | |
| | 11 | 三波川帯 Sanbagawa Belt | | |

中央構造線
Median Tectonic Line

糸魚川ー静岡構造線
Itoigawa-Shizuoka Tectonic Line

**付図5**　（相澤泰造氏からの提供）

## E. 数字・記号・数式の読み方

### Decimals and Fractions
### 小数と分数

| | |
|---|---|
| 12.503 | twelve point five zero three |
| 0.57 | zero point five seven |
| $\dfrac{1}{2}$ | one half |
| $\dfrac{1}{3}$ | one third |
| $\dfrac{1}{4}$ | one quarter |
| $\dfrac{1}{10}$ | one tenth |
| $5\dfrac{3}{5}$ | five and three fifths |
| $\dfrac{x}{y}$ | $x$ over $y$ ($x$ by $y$) |

### Index of exponential and Functions
### 指数・関数

| | |
|---|---|
| $x^2$ | $x$ squared |
| $x^3$ | $x$ cubed |
| $x^n$ | $x$ to the power $n$ |
| $x^{-n}$ | $x$ to the power minus $n$ |
| $\sqrt{x}$ | the square root of $x$ |
| $\sqrt[3]{x}$ | the cube root of $x$ |
| $\sqrt[n]{x}$ | the $n$-th root of $x$ |
| $\exp x$ | exponential $x$ |
| $\log_n x$ | log $x$ to the base $n$ |
| $\ln x$ | natural log $x$ |

付　　　録　　149

**superscripts and subscripts**
**上つき・下つき**

| | |
|---|---|
| $A_n$ | $A$ subscript $n$ |
| $A^n$ | $A$ superscript $n$ |
| $A'$ | $A$ prime (dash) |
| $A^*$ | $A$ asterisk ($A$ star) |
| $\overline{A}$ | $A$ bar |

**Operations and Operators**
**演算**

| | |
|---|---|
| $x+y=z$ | $x$ plus $y$ is equal to $z$ |
| $x-y=z$ | $x$ minus $y$ is equal to $z$ |
| $x \times y=z$ | $x$ times $y$ is equal to $z$ |
| $x \div y=z$ | $x$ divided by $y$ is equal to $z$ |
| $x^2-y^2=(x+y)(x-y)$ | $x$ squared minus $y$ squared is equal to the quantity $x$ plus $y$ times the quantity $x$ minus $y$ |

**Differentials and Derivatives**
**微分と導関数**

| | |
|---|---|
| $f(x)=\sin x$ | function of $x$ is equal to sine $x$ |
| $df=\cos x\, dx$ | differential of $f$ is equal to cosine $x$ $dx$ |
| $\dfrac{df}{dx}=\cos x$ | derivative of $f$ with respect to $x$ is equal to cosine $x$ |
| $\dfrac{\partial u}{\partial x}=y\cos(xy)$ | partial derivative of $u$ with respect to $x$ is equal to $y$ times cosine $xy$ |

$\dfrac{d^2y}{dx^2} = -\sin x$    second derivative of $y$ with respect to $x$ is equal to minus sine $x$

$\dfrac{\partial^2 u}{\partial x^2}$    second partial derivative of $u$ with respect to $x$

$\dfrac{\partial^2 z}{\partial x \partial y}$    second partial derivative of $z$ with respect to $x$ and $y$

$\dfrac{d^n y}{dx^n}$    $n$ th derivative of $y$ with respect to $x$

**Integrals 積分**

$\sum\limits_{i=1}^{n} a_i$    Summation of $a$ subscript $i$ where $i$ is equal to 1 to $n$

$\int k f(x)\,dx = k \int f(x)\,dx$    Integration of a constant, $k$ times function of $x$ with respect to $x$ is equal to $k$ times integration of function of $x$ with respect to $x$

$\int_a^b f(x)\,dx$    Integration of function of $x$ with respect to $x$ from $a$ to $b$

$\iint f(x, y)\,dxdy$    Double integration of function of $(x, y)$

$\int_c^d \int_a^b f(x, y)\,dxdy$    Integration of function of $(x, y)$ with respect to $x$ from $a$ to $b$ and with respect to $y$ from $c$ to $d$

**Symbols 記号**

$x_k \to 0$ as $k \to \infty$    $x$ subscript $k$ tends to 0 as $k$ tends to infinity

$F(x) \equiv 0$    Function of $x$ is identical to 0

$\hat{x} \fallingdotseq x$    $x$ hat is approximately equal to $x$

$\bar{x} \neq x^*$    $x$ bar is not equal to $x$ star (or $x$ asterisk)

$x > y$    $x$ is greater than $y$

$\tilde{x} < x'$    $x$ tilde is less than $x$ prime

$x < y$    $x$ is less than $y$

$x \geqq y$    $x$ is greater than or equal to $y$

$x \leqq y$    $x$ is less than or equal to $y$

$\lim_{k \to \infty} \dfrac{1}{k}(a^k + b^{\frac{2}{3}})$    the limit as $k$ goes to infinity of the quantity one over $k$ times the quantity $a$ to the $k$ th power plus $b$ to the two-thirds power

## F. ギリシャ文字

### Greek Alphabet

| | | | | | | | | |
|---|---|---|---|---|---|---|---|---|
| A | $\alpha$ | alpha | (アルファ) | N | $\nu$ | nu | (ニュー) |
| B | $\beta$ | beta | (ベータ) | Ξ | $\xi$ | xi | (クサイ, クシー) |
| Γ | $\gamma$ | gamma | (ガンマ) | O | $o$ | omicron | (オミクロン) |
| Δ | $\delta$ | delta | (デルタ) | Π | $\pi$ | pi | (パイ) |
| E | $\varepsilon$ | epsilon | (イプシロン) | P | $\rho$ | rho | (ロー) |
| Z | $\zeta$ | zeta | (ジータ) | Σ | $\sigma$ | sigma | (シグマ) |
| H | $\eta$ | eta | (イータ) | T | $\tau$ | tau | (タウ) |
| Θ | $\theta$ | theta | (シータ) | Υ | $\upsilon$ | upsilon | (ウプシロン) |
| I | $\iota$ | iota | (イオタ) | Φ | $\varphi, \phi$ | phi | (ファイ) |
| K | $\varkappa$ | kappa | (カッパ) | X | $\chi$ | chi | (キー, カイ) |
| Λ | $\lambda$ | lambda | (ラムダ) | Ψ | $\psi, \phi$ | psi | (プサイ) |
| M | $\mu$ | mu | (ミュー) | Ω | $\omega$ | omega | (オメガ) |

# 参　考　文　献

1) 編集委員会編：土質試験―基本と手引き―，地盤工学会（2001）
2) 編集委員会編：土質試験の方法と解析，地盤工学会（2000）
3) 大根義男：実務者のための土質力学，技報堂出版（2006）
4) 松岡　元：土質力学，基礎土木工学シリーズ，森北出版（1999）
5) 今井五郎：わかりやすい土の力学，鹿島出版会（1983）
6) 山口柏樹：土質力学―講義と演習―全改訂版，技報堂出版（1984）
7) 石原研而：土質力学，丸善（1988）
8) 河上房義：土質力学，森北出版（1992）
9) 伊藤　実：よくわかる土質力学例題集，工学出版（1993）
10) 澤　孝平・渡辺康二・沖村　孝・青木一男・佐野博昭：地盤工学，建設工学シリーズ，森北出版（1999）
11) 松田敬一郎ほか：土壌学，文永堂出版（1989）
12) 赤井浩一：土質力学，朝倉土木工学講座，朝倉書店（1966）
13) 酒井俊典・勝山邦久・Md. Zakaria Hossain・Laura J. Pyrak-Nolte：英語で学ぶ土質力学（1），コロナ社（2010）
14) 勝山邦久・酒井俊典・H. P. Rossmanith：英語で学ぶ構造力学，コロナ社（2006）
15) Arora, K. R.：Soil Mechanics and Foundation Engineering, Standard Publishers Distributors, Delhi, India (1989)
16) Banerjee, P. K. and Butterfield, R.：Development in Soil Mechanics and Foundation Engineering-2, Elsevier Applied Science Publishers Ltd. England (1985)
17) Bishop, A. W.：The Use of The Slip Circle in The Stability Analysis of Slopes, Geotechnique, **5**, 1 (1955)
18) Coulomb, C. A.：Essai Sur Une Application Des Regles De Maximis Et Minimus A Quelques Problems De Statique Relatifs A Parchitecture, Memotre De La Mathematique Et De Physique, **7**, Paris (1773), De L, Imprimeric Royale (1776)

19) Das, B. M. : Advanced Soil Mechanics, Hemisphere Publishing Corporation, McGraw-Hill Book Company, USA (1983)
20) Fellenius, W. : Erdstatische Berechnungen Mit Reibung Und Kohesion Adhesion, Und Unter Annihme Kreiszylindrischer Gleit Flachen. Rev. Ed. Berlin, Ernst (1939)
21) Mayerhof, G. C. : Ultimate Bearing Capacity of Foundations, Geotechnique, **2** (1951)
22) Prandtl, L. : Uber Die Eindringungsfestigkei (Harte) Plastischer Baustoffe Und Die Festigkeit Von Schneiden, Zeitschrift Fur Angewandte Mathematik Und Mechanik, **1** (1921)
23) Rankine, W. J. M. : On the Stability of Loose Earth, Phill Trans. Royal Soc., **47**, London (1857)
24) Rankine, W. J. M. : A Manual of Applied Mechanics, Charles Griffin & Co. London (1885)
25) Scott, C. R. : An Introduction to Soil Mechanics & Foundations, Applied Science Publishers Ltd. London (1980)
26) Scott, C. R. : Developments in Soil Mechanics, Applied Science Publishers Ltd. London (1978)
27) Scott, R. F. : Principles of Soil Mechanics, Addison-Wesley Publishing Company, Inc. USA (1963)
28) Terzaghi, K. : Theoretical Soil Mechanics, Wiley, New York (1943)
29) Yong, R. N. and Warkentin, B. P. : Introduction to Soil Behavior, The Macmillan Company, New York, USA (1966)

# 英和索引

## A

accumulation ……………………57
　堆　積
active condition ………………………84
　主働状態
active earth pressure ……………90, 98
　主働土圧
active earth pressure coefficient ……112
　主働土圧係数
active wedge domain ………………117
　主働くさび領域
adsorption water ………………………41
　吸着水
adsorption water layer ………………40
　吸着水層
angle of internal friction
　………………………38, 58, 114, 118
　内部摩擦角
apparatus ………………………………58
　試験機

## B

balance of forces ………………98, 128
　力の釣合い
bearing capacity ……………………106
　支持力
bearing capacity equation ……………114
　支持力公式
Bishop ………………………………140
　ビショップ

block ……………………………128
　ブロック

## C

CD ………………………………74, 78
center of gravity ……………………22
　重　心
circular shape ………………………118
　円　形
circular slip surface ………………136
　円弧すべり
clay ……………………………………36
　粘　土
clayey soil ……………………………76
　粘性土
clay particle …………………………41
　粘土粒子
coarse particle ………………………36
　粗粒分
coefficient of active earth pressure
　……………………………………86, 100
　主働土圧係数
coefficient of earth pressure ………89
　土圧係数
coefficient of earth pressure at rest …84
　静止土圧係数
coefficient of passive earth pressure
　……………………………………86, 102
　受働土圧係数
coefficient of permeability …………42, 76
　透水係数

英　和　索　引　*155*

coefficient of bearing capacity ……… 118
　　支持力係数
cohesion ……………………… 40, 58, 114
　　粘着力
collapse ……………………………… 126
　　崩　壊
compaction pile ……………………… 122
　　締固め杭
compression …………………………… 62
　　圧　縮
compressive shear failure …………… 90
　　圧縮せん断破壊
compressive strain …………………… 63
　　圧縮ひずみ
compressive stress …………………… 62
　　圧縮応力
confine ………………………………… 66
　　拘　束
conjugate relationship …………… 22, 24
　　共役関係
consolidated and drained …………… 74
　　圧密・排水
consolidated and drained test ……… 78
　　圧密・排水試験
consolidated and undrained ………… 74
　　圧密・非排水
consolidated and undrained test …… 76
　　圧密・非排水試験
consolidation …………………… 58, 72
　　圧　密
consolidation condition ……………… 66
　　圧密条件
consolidation history ………………… 56
　　圧密履歴
constant pressure shear test ………… 60
　　定圧試験
constant volume shear test ………… 60
　　定体積試験
contact area …………………………… 36
　　接触面積

continuous …………………………… 118
　　連　続
continuous footing ………………… 118
　　連続フーチング
controlled pressure ………………… 67
　　拘束圧
Coulomb's earth pressure …………… 96
　　クーロン土圧
Coulomb's earth pressure coefficient
　　……………………………………… 102
　　クーロンの土圧係数
Coulomb's failure criterion
　　………………… 36, 40, 46, 58, 130, 138
　　クーロンの破壊基準
cross-sectional area ………………… 143
　　断面積
CU ………………………………… 74, 76
$\overline{\text{CU}}$ ………………………………… 74, 76

## D

decrease in volume …………………… 55
　　体積減少
deep foundation ………………… 114, 120
　　深い基礎
deformaton …………………………… 35
　　変　形
dense condition ……………………… 54
　　密詰め
dense sand ……………………… 55, 56
　　密詰め砂
dilatancy ……………………………… 54
　　ダイレイタンシー
dilatancy property …………………… 54
　　ダイレイタンシー特性
direction of force …………………… 36
　　力の作用方向
direct shear testing apparatus ……… 58
　　一面せん断試験

disturbed-soil ················66, 70
　練返した土
drainage ·····················72
　排　水
drainage condition ··············66
　排水条件
drained water ···············67, 72
　排　水

## E

$E_{50}$ ·······················64
earth pressure ··················84
　土　圧
effective stress
　················42, 72, 74, 76, 82, 132, 138
　有効応力
elastic ························8
　弾　性
elastic body ····················6
　弾性体
elasto-plastic body ···············8
　弾塑性体
encroachment ··················57
　侵　食
end bearing pile ···············120
　先端支持杭
equilibrium ················98, 128
　釣合い
equilibrium of moment ············22
　モーメントの釣合い
excess pore-water pressure
　··················76, 78, 42, 72
　過剰間隙水圧
expansion shear failure ···········90
　伸張せん断破壊

## F

failure condition ···············96
　破壊条件

failure envelope ···········46, 66, 74
　破壊包絡線
failure stress circle ··············46
　破壊応力円
Fellenius method ··············136
　フェレニウス法
fine particle ···················36
　細粒分
footing ·····················106
　フーチング
foundation ··················106
　基　礎
frictional force ·················38
　摩擦抵抗力
frictional resistance ··············38
　摩擦抵抗
frictional resistance ············114
　摩擦抵抗力
friction pile ·················120
　摩擦杭
$F_s$ ·······················136

## G

general shear failure ············106
　全般せん断破壊
gravel ····················36, 76
　礫
gravitational force ·············126
　重　力
ground surface ···············117
　地表面
groups of piling ···············122
　群　杭

## H

Hooke's law ···················8
　フックの法則

英　和　索　引　*157*

## I

increase in volume ·····················55
　体積増加
infinitesimal element ·····················22
　微小要素
initial void ratio ························54
　初期間隙比

## K

$K_0$ ····················································84
$K_a$ ·············································86, 100
$K_p$ ·············································86, 102

## L

limit equilibrium method ···············126
　極限釣合い法
load ·········································35, 107
　荷　重
load-settlement relationship ············106
　荷重-沈下関係
local shear failure ······················106
　局所せん断破壊
longitudinal strain ························5
　縦ひずみ
long-term consolidation problem ······80
　長期安定問題
loose condition ·····················54
　緩詰め
loose sand ·····························55, 56
　緩詰め砂

## M

material property ·····················38
　物　性
maximum compressive strength ······62
　最大の圧縮応力
maximum principal shear stress ······28
　最大主せん断応力

maximum principal stress ··········18, 28
　最大主応力
minimum principal stress ·········18, 28
　最小主応力
modulus of deformation ···············64
　変形係数
Mohr-Coulomb's failure criterion
　·············································46, 48
　モール・クーロンの破壊基準
Mohr's stress circle ······14, 16, 24, 28, 66
　モールの応力円
movement of soil particle ···············87
　土粒子の移動
moving soil block ·····················129
　移動土塊

## N

negative friction ·····················122
　ネガティブフリクション
negatively charged ion ···············40
　負電荷
normal load ·····························44, 58
　垂直荷重
normally consolidated ···············56
　正規圧密
normal stress ·················2, 24, 34, 44
　垂直応力

## O

over-consolidated ·····················56
　過圧密

## P

$p$ ····················································48
particle crushing ·····················35
　粒子破砕
passive condition ·····················84, 86
　受働状態
passive earth pressure ···············90
　受働土圧

passive earth pressure coefficient ⋯112
   受働土圧係数
peak strength ⋯⋯⋯⋯⋯⋯⋯⋯⋯⋯56
   ピーク強度
piling ⋯⋯⋯⋯⋯⋯⋯⋯⋯⋯⋯⋯⋯106
   杭
plastic domain ⋯⋯⋯⋯⋯⋯⋯⋯110
   塑性領域
Poisson's ratio ⋯⋯⋯⋯⋯⋯⋯⋯⋯4
   ポアソン比
pore water pressure ⋯⋯⋯⋯42, 67, 74
   間隙水圧
($p$-$q$) coordinate ⋯⋯⋯⋯⋯⋯⋯50
   $p$-$q$ 座標
$p$-$q$ plot ⋯⋯⋯⋯⋯⋯⋯⋯⋯⋯⋯48
   $p$-$q$ プロット
principal stress ⋯⋯⋯⋯⋯⋯14, 46, 90
   主応力
principal surface ⋯⋯⋯⋯⋯⋯⋯⋯30
   主応力面

## Q

$q$ ⋯⋯⋯⋯⋯⋯⋯⋯⋯⋯⋯⋯⋯⋯50

## R

Rankine ⋯⋯⋯⋯⋯⋯⋯⋯⋯⋯⋯108
   ランキン
Rankine's active earth pressure
   coefficient ⋯⋯⋯⋯⋯⋯⋯⋯114
   ランキンの受働土圧係数
Rankine's bearing capacity formula
   ⋯⋯⋯⋯⋯⋯⋯⋯⋯⋯⋯⋯⋯114
   ランキンの支持力公式
Rankine's earth pressure coefficient
   ⋯⋯⋯⋯⋯⋯⋯⋯⋯⋯⋯⋯90, 94
   ランキンの土圧係数
rectangular shape ⋯⋯⋯⋯⋯⋯⋯118
   長方形
residual strain ⋯⋯⋯⋯⋯⋯⋯⋯⋯8
   残留ひずみ

residual strength ⋯⋯⋯⋯⋯⋯⋯56
   残留強度
resisting moment ⋯⋯⋯⋯⋯⋯⋯136
   抵抗モーメント
rest condition ⋯⋯⋯⋯⋯⋯⋯⋯⋯84
   静止状態
rigid-plastic body ⋯⋯⋯⋯⋯⋯⋯8
   剛塑性体
roughness ⋯⋯⋯⋯⋯⋯⋯⋯⋯⋯108
   粗い

## S

safety factor ⋯⋯⋯⋯⋯⋯⋯⋯⋯130
   安全率
sand ⋯⋯⋯⋯⋯⋯⋯⋯⋯⋯⋯36, 76
   砂
saturated ⋯⋯⋯⋯⋯⋯⋯⋯⋯⋯76
   飽和
saturated condition ⋯⋯⋯⋯⋯⋯130
   飽和状態
self weight ⋯⋯⋯⋯⋯⋯⋯⋯⋯143
   自重
sensitivity ratio ⋯⋯⋯⋯⋯⋯⋯⋯66
   鋭敏比
settlement ⋯⋯⋯⋯⋯⋯⋯⋯⋯106
   沈下
shallow foundation ⋯⋯⋯⋯⋯⋯114
   浅い基礎
shape factor ⋯⋯⋯⋯⋯⋯⋯⋯⋯120
   形状係数
shear deformation ⋯⋯⋯⋯⋯⋯54
   せん断変形
shear displacement ⋯⋯⋯⋯56, 60
   せん断変位
shear force ⋯⋯⋯⋯⋯⋯⋯⋯6, 44
   せん断力
shear modulus ⋯⋯⋯⋯⋯⋯⋯⋯8
   せん断係数
shear strain ⋯⋯⋯⋯⋯⋯⋯⋯⋯6
   せん断ひずみ

shear strength ……………34, 58, 76
　せん断強度
shear stress …………6, 22, 24, 34, 44, 60
　せん断応力
shear surface …………14, 34, 35, 60, 86
　せん断面
shear testing apparatus ……………58
　せん断試験機
short-term stability problem …………76
　短期安定問題
silt …………………………………36
　シルト
simplified method ………………136
　簡便法
single pile ………………………122
　単杭
size of the particle ………………36
　粒径
slice ………………………………136
　スライス
slice method ……………………136
　分割法
sliding moment …………………136
　滑動モーメント
slip surface
　………34, 35, 86, 87, 98, 106, 126, 136
　すべり面
slope ……………………………126
　斜面
smoothness ……………………108
　滑らか
soil particle ………………………35
　土粒子
soil specimen ……………………58
　土供試体
specimen …………………………60
　供試体
square shape ……………………118
　正方形
$S_t$ …………………………………66

stability analysis ………………136
　安定解析
strain ………………………2, 4, 88, 89
　ひずみ
strain-hardening …………………9
　ひずみ硬化
strain-softening …………………10, 56
　ひずみ軟化
stress ……………………………2
　応力
stress circle at failure ……………46
　破壊応力円
stress history ……………………50
　応力履歴
stress path ………………50, 51, 60
　応力経路
stress-strain curve ………………62
　応力-ひずみ曲線
stress-strain relationship …………8
　応力-ひずみ関係
surcharge ………………………114
　サーチャージ

# T

tensile ……………………………2
　引張り
Terzaghi's bearing capacity formula
　………………………………116
　テルツァーギの支持力公式
total stress ………………42, 72, 76, 82
　全応力
transition domain ………………117
　遷移領域
transverse strain …………………5
　横ひずみ
triaxial compressive strength testing
　apparatus ……………………58
　三軸圧縮試験機

## U

ultimate bearing capacity ········108, 116
　極限支持力
unconfined ·································62
　拘束のない
unconfined compressive strength ······62
　一軸圧縮強さ
unconfined compressive strength test
　······································62, 74
　一軸圧縮試験
unconfined compressive strength testing
　apparatus ································58
　一軸圧縮試験機
unconsolidated and undrained ········74
　非圧密・非排水
unconsolidated and undrained test ···74
　非圧密・非排水試験
undisturbed-soil ····················66, 70
　乱さない土
unloading ··································8
　除荷
UU ········································74

## V

vertical strain ····························56
　垂直ひずみ
vertical wall ······························84
　垂直壁

viscosity ··································40
　粘　性
void ratio ·························72, 74, 76
　間隙比
volume change ·····················67, 78
　体積変化
volumetric strain ····················56, 58
　体積ひずみ

## W

water pressure ·················84, 138
　水　圧
wedge ······························108, 109
　くさび

## Y

yield load ······························107
　降伏荷重
Young's modulus ·························8
　弾性係数

## 【その他】

$\sigma_{ha}$ ·····································90
$\sigma_{hp}$ ·····································90
$\tau$ axis ··································60
　$\tau$　軸

―― 著者略歴 ――

**酒井　俊典**（さかい　としのり）
1981 年　愛媛大学農学部農業工学科卒業
1989 年　東京大学大学院農学系研究科博士課程修了（農業工学専攻），農学博士
1989 年　愛媛大学助手
1996 年　英国・グラスゴー大学客員研究員
1996 年　愛媛大学助教授
2002 年　英国・ブリストル大学客員研究員
2006 年　三重大学大学院教授
　　　　現在に至る

**勝山　邦久**（かつやま　くにひさ）
1968 年　京都大学工学部資源工学科卒業
1973 年　京都大学工学研究科博士課程修了
1973 年　通商産業省・工業技術院公害資源研究所（現経済産業省・産業技術総合研究所），
　　　　工学博士（京都大学）
1978 年
〜80 年　カナダ・CANMET 客員研究員
1996 年　工業技術院資源環境技術総合研究所部長
2000 年　愛媛大学教授
2010 年　愛媛大学名誉教授

**Md. Zakaria Hossain**
1990 年　バングラデシュ農業大学農業工学部卒業
1995 年　京都大学大学院農学研究科修士課程修了（農業工学専攻）
1998 年　京都大学大学院農学研究科博士後期課程修了（地域環境科学専攻），博士（農学）
1998 年　日本学術振興会外国人特別研究員(神戸大学)
1999 年　三重大学助手
2000 年　カナダ・ブリティッシュコロンビア大学客員研究員
2003 年　三重大学講師
2008 年　三重大学大学院准教授
　　　　現在に至る

**Laura J. Pyrak-Nolte**
1988 年　米国・カリフォルニア大学バークレー校修了，Ph.D.
1992 年　米国・ノートルダム大学助教授
1997 年　米国・パデュー大学教授
　　　　現在に至る
1995 年　国際岩の力学学会シュルンベルジュ賞受賞

英語で学ぶ 土質力学 (2) ― 力学的性質編 ―
Soil Mechanics (2) ― Mechanical Properties ―
　　　　　　　　　　© Sakai, Katsuyama, Hossain, Pyrak-Nolte　2010

2010年 9 月 22 日　初版第 1 刷発行　　　　　　　　　　★

|検印省略| 著　者 | 酒　井　俊　典 |
| | | 勝　山　邦　久 |
| | | Md. Zakaria Hossain |
| | | Laura J. Pyrak-Nolte |
| | 発 行 者 | 株式会社　コロナ社 |
| | | 代 表 者　牛来真也 |
| | 印 刷 所 | 壮光舎印刷株式会社 |

112-0011　東京都文京区千石 4-46-10
発行所　株式会社 コロナ社
CORONA PUBLISHING CO., LTD.
Tokyo　Japan
振替 00140-8-14844・電話 (03) 3941-3131 (代)
ホームページ http://www.coronasha.co.jp

ISBN 978-4-339-05225-1　　（高橋）　　（製本：グリーン）
Printed in Japan

無断複写・転載を禁ずる
落丁・乱丁本はお取替えいたします

# 環境・都市システム系教科書シリーズ

(各巻A5判，14.のみB5判)

■編集委員長　澤　孝平
■幹　　　事　角田　忍
■編集委員　　荻野　弘・奥村充司・川合　茂
　　　　　　　嵯峨　晃・西澤辰男

| 配本順 | | | 頁 | 定価 |
|---|---|---|---|---|
| 1. (16回) | シビルエンジニアリングの第一歩 | 澤 孝平・嵯峨 晃<br>川合 茂・角田 忍<br>荻野 弘・奥村充司 共著<br>西澤辰男 | 176 | 2415円 |
| 2. (1回) | コンクリート構造 | 角田　忍<br>竹村和夫 共著 | 186 | 2310円 |
| 3. (2回) | 土質工学 | 赤木知之・吉村優治<br>上 俊二・小堀慈久 共著<br>伊東 孝 | 238 | 2940円 |
| 4. (3回) | 構造力学Ⅰ | 嵯峨 晃・武田八郎<br>原 隆・勇 秀憲 共著 | 244 | 3150円 |
| 5. (7回) | 構造力学Ⅱ | 嵯峨 晃・武田八郎<br>原 隆・勇 秀憲 共著 | 192 | 2415円 |
| 6. (4回) | 河川工学 | 川合 茂・和田 清<br>神田佳一・鈴木正人 共著 | 208 | 2625円 |
| 7. (5回) | 水理学 | 日下部重幸・檀 和秀<br>湯城豊勝 共著 | 200 | 2730円 |
| 8. (6回) | 建設材料 | 中嶋清実・角田 忍<br>菅原 隆 共著 | 190 | 2415円 |
| 9. (8回) | 海岸工学 | 平山秀夫・辻本剛三<br>島田富美男・本田尚正 共著 | 204 | 2625円 |
| 10. (9回) | 施工管理学 | 友久誠司<br>竹下治之 共著 | 240 | 3045円 |
| 11. (10回) | 測量学Ⅰ | 堤　　隆 著 | 182 | 2415円 |
| 12. (12回) | 測量学Ⅱ | 岡林 巧・堤 隆<br>山田貴浩 共著 | 214 | 2940円 |
| 13. (11回) | 景観デザイン<br>―総合的な空間のデザインをめざして― | 市坪 誠・小川総一郎<br>谷平 考・砂本文彦 共著<br>溝上裕二 | 222 | 3045円 |
| 14. (13回) | 情報処理入門 | 西澤辰男・長岡健一<br>廣瀬康之・豊田 剛 共著 | 168 | 2730円 |
| 15. (14回) | 鋼構造学 | 原 隆・山口隆司<br>北原武嗣・和多田康男 共著 | 224 | 2940円 |
| 16. (15回) | 都市計画 | 平田登基男・亀野辰三<br>宮腰和弘・武井幸久 共著<br>内田一平 | 204 | 2625円 |
| 17. (17回) | 環境衛生工学 | 奥村充司<br>大久保孝樹 共著 | 238 | 3150円 |
| 18. (18回) | 交通システム工学 | 大橋健一・柳澤吉保<br>髙岸節夫・佐々木恵一<br>日野 智・折田仁典 共著<br>宮腰和弘・西澤辰男 | 224 | 2940円 |

以下続刊

防災工学　　　　　渕田・塩野・檀<br>疋田・吉村　共著　　環境保全工学　和田・奥村共著

建設システム計画　荻野・大橋・野田<br>西澤・鈴木　共著

定価は本体価格+税5%です。
定価は変更されることがありますのでご了承下さい。

図書目録進呈◆

# 土木系 大学講義シリーズ

(各巻A5判，欠番は品切です)

■編集委員長　伊藤　學
■編集委員　青木徹彦・今井五郎・内山久雄・西谷隆亘
　　　　　　榛沢芳雄・茂庭竹生・山﨑　淳

| 配本順 | | | 頁 | 定価 |
|---|---|---|---|---|
| 2.（4回） | 土木応用数学 | 北田俊行著 | 236 | 2835円 |
| 3.（27回） | 測量学 | 内山久雄著 | 206 | 2835円 |
| 4.（21回） | 地盤地質学 | 今井・福江・足立 共著 | 186 | 2625円 |
| 5.（3回） | 構造力学 | 青木徹彦著 | 340 | 3465円 |
| 6.（6回） | 水理学 | 鮏川　登著 | 256 | 3045円 |
| 7.（23回） | 土質力学 | 日下部　治著 | 280 | 3465円 |
| 8.（19回） | 土木材料学（改訂版） | 三浦　尚著 | 224 | 2940円 |
| 9.（13回） | 土木計画学 | 川北・榛沢編著 | 256 | 3150円 |
| 11.（17回） | 改訂 鋼構造学 | 伊藤　學著 | 260 | 3360円 |
| 13.（7回） | 海岸工学 | 服部昌太郎著 | 244 | 2625円 |
| 14.（25回） | 改訂 上下水道工学 | 茂庭竹生著 | 240 | 3045円 |
| 15.（11回） | 地盤工学 | 海野・垂水編著 | 250 | 2940円 |
| 16.（12回） | 交通工学 | 大蔵　泉著 | 254 | 3150円 |
| 17.（26回） | 都市計画（三訂版） | 新谷・髙橋・岸井 共著 | 190 | 2730円 |
| 18.（24回） | 新版 橋梁工学（増補） | 泉・近藤共著 | 324 | 3990円 |
| 20.（9回） | エネルギー施設工学 | 狩野・石井共著 | 164 | 1890円 |
| 21.（15回） | 建設マネジメント | 馬場敬三著 | 230 | 2940円 |
| 22.（22回） | 応用振動学 | 山田・米田共著 | 202 | 2835円 |

以下続刊

10.　コンクリート構造学　山﨑　淳著　　12.　河川工学　西谷隆亘著
19.　水環境システム　大垣真一郎 他著

定価は本体価格＋税5％です。
定価は変更されることがありますのでご了承下さい。

図書目録進呈◆

# 新編土木工学講座

(各巻A5判，欠番は品切です)

■全国高専土木工学会編
■編集委員長　近藤泰夫

| 配本順 | | 書名 | 著者 | 頁 | 定価 |
|---|---|---|---|---|---|
| 1. | (3回) | 土木応用数学 | 近藤・江崎共著 | 322 | 3675円 |
| 2. | (21回) | 土木情報処理 | 杉山・錦雄栗木・譲共著 | 282 | 2940円 |
| 4. | (22回) | 土木工学概論 | 長谷川博他著 | 220 | 2310円 |
| 6. | (29回) | 測量（1）(新訂版) | 長谷川・植田大木共著 | 270 | 2730円 |
| 7. | (30回) | 測量（2）(新訂版) | 小川・植田大木共著 | 304 | 3150円 |
| 8. | (27回) | 新版 土木材料学 | 近藤・岸角田本共著 | 312 | 3465円 |
| 9. | (2回) | 構造力学（1）<br>―静定編― | 宮原・高端共著 | 310 | 3150円 |
| 11. | (11回) | 新版 土質工学 | 中野・小杉山山共著 | 240 | 2835円 |
| 12. | (9回) | 水理学 | 細井・杉山共著 | 360 | 3150円 |
| 13. | (25回) | 新版 鉄筋コンクリート工学 | 近藤・岸角田本共著 | 310 | 3570円 |
| 14. | (26回) | 新版 橋工学 | 高端・向山久保田共著 | 276 | 3570円 |
| 15. | (19回) | 土木施工法 | 伊丹・片山後藤・原島共著 | 300 | 3045円 |
| 16. | (10回) | 港湾および海岸工学 | 菅野・寺西堀口・佐藤共著 | 276 | 3150円 |
| 17. | (17回) | 改訂 道路工学 | 安孫子・澤共著 | 336 | 3150円 |
| 18. | (13回) | 鉄道工学 | 宮原・雨宮共著 | 216 | 2625円 |
| 19. | (28回) | 新 地域および都市計画（改訂版） | 岡崎・高岸大橋・竹内共著 | 218 | 2835円 |
| 21. | (16回) | 河川および水資源工学 | 渋谷・大同共著 | 338 | 3570円 |
| 22. | (15回) | 建築学概論 | 橋本・渋谷大沢・谷本共著 | 278 | 3045円 |
| 23. | (23回) | 土木耐震工学 | 狩俣・音田荒川共著 | 202 | 2625円 |

定価は本体価格+税5％です。
定価は変更されることがありますのでご了承下さい。

図書目録進呈◆

# 地球環境のための技術としくみシリーズ

（各巻A5判）

コロナ社創立75周年記念出版　〔創立1927年〕

■編集委員長　松井三郎
■編集委員　　小林正美・松岡　譲・盛岡　通・森澤眞輔

|配本順| | |頁|定価|
|---|---|---|---|---|
|1.（1回）|今なぜ地球環境なのか|松井三郎編著|230|3360円|
| |松下和夫・中村正久・髙橋一生・青山俊介・嘉田良平 共著| | | |
|2.（6回）|生活水資源の循環技術|森澤眞輔編著|304|4410円|
| |松井三郎・細井由彦・伊藤禎彦・花木啓祐　　　　　共著| | | |
| |荒巻俊也・国包章一・山村尊房| | | |
|3.（3回）|地球水資源の管理技術|森澤眞輔編著|292|4200円|
| |松岡　譲・髙橋　潔・津野　洋・古城方和　　　　　共著| | | |
| |楠田哲也・三村信男・池淵周一| | | |
|4.（2回）|土壌圏の管理技術|森澤眞輔編著|240|3570円|
| |米田　稔・平田健正・村上雅博 共著| | | |
|5.|資源循環型社会の技術システム|盛岡　通編著| | |
| |河村清史・吉田　登・藤田　壯・花嶋正孝　　　　　共著| | | |
| |宮脇健太郎・後藤敏彦・東海明宏| | | |
|6.（7回）|エネルギーと環境の技術開発|松岡　譲編著|262|3780円|
| |森　俊介・槌屋治紀・藤井康正 共著| | | |
|7.|大気環境の技術とその展開|松岡　譲編著| | |
| |森口祐一・島田幸司・牧野尚夫・白井裕三・甲斐沼美紀子 共著| | | |
|8.（4回）|木造都市の設計技術| |282|4200円|
| |小林正美・竹内典之・髙橋康夫・山岸常人　　　　　共著| | | |
| |外山　義・井上由起子・菅野正広・鉾井修一| | | |
| |吉田治典・鈴木祥之・渡邉史夫・高松　伸| | | |
|9.|環境調和型交通の技術システム|盛岡　通編著| | |
| |新田保次・鹿島　茂・岩井信夫・中川　大　　　　　共著| | | |
| |細川恭史・林　良嗣・花岡伸也・青山吉隆| | | |
|10.|都市の環境計画の技術としくみ|盛岡　通編著| | |
| |神吉紀世子・室崎益輝・藤田　壯・島谷幸宏　　　　共著| | | |
| |福井弘道・野村康彦・世古一穂| | | |
|11.（5回）|地球環境保全の法としくみ|松井三郎編著|330|4620円|
| |岩間　徹・浅野直人・川勝健志・植田和弘　　　　　共著| | | |
| |倉阪秀史・岡島成行・平野　喬| | | |

定価は本体価格+税5％です。
定価は変更されることがありますのでご了承下さい。

図書目録進呈◆

# シリーズ　21世紀のエネルギー

(各巻A5判)

■(社)日本エネルギー学会編

| | | | 頁 | 定価 |
|---|---|---|---|---|
| 1. | **21世紀が危ない**<br>— 環境問題とエネルギー — | 小島紀徳著 | 144 | 1785円 |
| 2. | **エネルギーと国の役割**<br>— 地球温暖化時代の税制を考える — | 十市　勉<br>小川芳樹 共著<br>佐川直人 | 154 | 1785円 |
| 3. | **風と太陽と海**<br>— さわやかな自然エネルギー — | 牛山　泉他著 | 158 | 1995円 |
| 4. | **物質文明を超えて**<br>— 資源・環境革命の21世紀 — | 佐伯康治著 | 168 | 2100円 |
| 5. | **Cの科学と技術**<br>— 炭素材料の不思議 — | 白石　大谷<br>京谷・山田 共著 | 148 | 1785円 |
| 6. | **ごみゼロ社会は実現できるか** | 行西正雄<br>本哲生 共著<br>立田真文 | 142 | 1785円 |
| 7. | **太陽の恵みバイオマス**<br>— CO₂を出さないこれからのエネルギー — | 松村幸彦著 | 156 | 1890円 |
| 8. | **石油資源の行方**<br>— 石油資源はあとどれくらいあるのか — | JOGMEC調査部編 | 188 | 2415円 |
| 9. | **原子力の過去・現在・未来**<br>— 原子力の復権はあるか — | 山地憲治著 | 170 | 2100円 |

以下続刊

| | | | |
|---|---|---|---|
| 農のエネルギー | 小林・木谷・楠山<br>上野・近藤 共著 | 21世紀の太陽電池技術 | 荒川裕則著 |
| 太陽光発電の社会学 | 黒川浩助著 | キャパシタ<br>— これからの「電池ではない電池」 — | 直井勝彦著 |
| マルチガス削減<br>— エネルギー起源CO₂以外の温暖化要因を含めた総合対策 — | 黒沢敦志著 | 石炭資源の行方<br>— 21世紀の石炭資源開発技術 — | 島田荘平著 |
| バイオマスタウン | 森塚秀人他著 | | |

定価は本体価格+税5%です。
定価は変更されることがありますのでご了承下さい。

図書目録進呈◆

## 技術英語・学術論文書き方関連書籍

### 技術レポート作成と発表の基礎技法
野中謙一郎・渡邉力夫・島野健仁郎・京相雅樹・白木尚人 共著
A5／160頁／定価2,100円／並製

### マスターしておきたい 技術英語の基本
Richard Cowell・佘　錦華 共著
A5／190頁／定価2,520円／並製

### 科学英語の書き方とプレゼンテーション
日本機械学会 編／石田幸男 編著
A5／184頁／定価2,310円／並製

### 続 科学英語の書き方とプレゼンテーション
－スライド・スピーチ・メールの実際－
日本機械学会 編／石田幸男 編著
A5／176頁／定価2,310円／並製

### いざ国際舞台へ！
### 理工系英語論文と口頭発表の実際
富山真知子・富山　健 共著
A5／176頁／定価2,310円／並製

### 知的な科学・技術文章の書き方
－実験リポート作成から学術論文構築まで－
中島利勝・塚本真也 共著
A5／244頁／定価1,995円／並製　日本工学教育協会賞（著作賞）受賞

### 知的な科学・技術文章の徹底演習
塚本真也 著
A5／206頁／定価1,890円／並製　工学教育賞（日本工学教育協会）受賞

### 科学技術英語論文の徹底添削
－ライティングレベルに対応した添削指導－
絹川麻理・塚本真也 共著
A5／200頁／定価2,520円／並製